Name that Flower

Name that Flower

The Identification of Flowering Plants

SECOND EDITION

Ian Clarke and Helen Lee

MELBOURNE UNIVERSITY PRESS

MELBOURNE UNIVERSITY PRESS
an imprint of Melbourne University Publishing Ltd
PO Box 1167, Carlton, Victoria 3053, Australia
mup-info@unimelb.edu.au
www.mup.com.au

First published 1987
Reprinted 1989, 1990, 1992, 1993, 1994 (twice), 1995, 1997, 1998, 2001
Second edition 2003
Text © Ian Clarke and Helen Lee 1987, 2003
Illustrations © Ian Clarke 2003
Design and typography © Melbourne University Publishing Ltd 2003

This book is copyright. Apart from any use permitted under the *Copyright Act 1968* and subsequent amendments, no part may be reproduced, stored in a retrieval system or transmitted by any means or process whatsoever without the prior written permission of the publishers.

Designed and typeset in Australia by Lauren Statham, Alice Graphics
Printed in Australia by BPA Print Group
Cover designed by Ruth Grüner
Cover illustration: *Chamelaucium* sp. (wax-flower), family MYRTACEAE; photograph by Ian Clarke

National Library of Australia Cataloguing-in-Publication entry

Clarke, Ian, 1950– .
    Name that flower: the indentification of flowering plants
    2nd ed.
    Bibliography.
    Includes index.
    ISBN 0 522 85060 X.
    1. Flowers – Identification. 2. Botany – Australia.
    I. Lee, Helen, 1925– . II. Title.
582.13012

Contents

Plates	ix
Acknowledgements	xii
Introduction	xiii
Biographical Note	xv
Symbols and Abbreviations	xvi

1 Getting Started — 1
- Background — 1
- Some essential information — 2
- Using this book — 4
- Notes on the illustrations — 4

2 The Structure of Flowers — 6
- Structure of a basic flower — 7
- Variation in floral structure — 7

3 Inflorescences—the arrangement of flowers on the plant — 31
- Solitary flowers — 31
- Racemose inflorescences — 31
- Cymose inflorescences — 31

4 Reproduction — 35
- Pollination — 35
- Fertilisation — 36
- Development of the seed — 37
- Development of the fruit — 37
- Germination — 39

Contents

5 Introduction to Plant Structure and Function 42
Cells 42
Stems 43
Leaves 43
Roots 50
The vascular system 50
Plant growth 51

6 Classification and Nomenclature—grouping and naming 52

7 The Process of Identification 58
Equipment 58
Choosing the flower to look at 59
Interpreting what you see 59
Using keys 61
Computer keys 68

8 Plant Families 70
Dicotyledons 70
CASUARINACEAE she-oaks 72
PROTEACEAE banksias, grevilleas, hakeas 72
LORANTHACEAE mistletoes 87
CHENOPODIACEAE saltbushes, samphires 88
RANUNCULACEAE buttercup family 92
LAURACEAE laurel family 94
DROSERACEAE sundews 95
TREMANDRACEAE 96
PITTOSPORACEAE pittosporum family 98
ROSACEAE rose family 103
MIMOSACEAE wattles 110
CAESALPINIACEAE cassias 117
FABACEAE pea family 120
RUTACEAE boronias, correas 129
EUPHORBIACEAE spurges 139
STACKHOUSIACEAE 142
RHAMNACEAE buckthorn family 144
STERCULIACEAE kurrajongs, paper-flowers 147
DILLENIACEAE guinea-flowers 150
THYMELAEACEAE rice-flowers 151
MYRTACEAE eucalypts, tea-trees, bottlebrushes 153

EPACRIDACEAE heath family	171
SOLANACEAE nightshades	183
LAMIACEAE mint-bushes	185
MYOPORACEAE myoporums and emubushes	188
GOODENIACEAE	189
BRUNONIACEAE	191
STYLIDIACEAE trigger-plants	191
ASTERACEAE daisy family	193
Monocotyledons	210
POACEAE grasses	211
XANTHORRHOEACEAE grass-trees, mat-rushes	218
LILIACEAE lily family	219
AMARYLLIDACEAE amaryllis family	224
IRIDACEAE iris family	228
ORCHIDACEAE orchid family	230
Appendix–Nomenclatural Changes since the First Edition	238
Bibliography, CD-Roms and Websites	240
Glossary	264
Index	282

Plates

Between pages 208 and 209

Plate 1   Floral Structure
Crassula multicava (crassula) Crassulaceae
Hibbertia empetrifolia (guinea-flower) Dilleniaceae
Boronia denticulata (boronia) Rutaceae
Tetratheca ciliata (pink-bells) Tremandraceae
Hypericum leschenaultii (St John's wort) Hypericaceae
Ranunculus sp. (buttercup) Ranunculaceae

Plate 2   Floral Structure
Grevillea rosmarinifolia (rosemary grevillea) Proteaceae
Grevillea sp. (grevillea) Proteaceae
Physalis viscosa (sticky ground-cherry) Solanaceae
Epacris impressa (common heath) Epacridaceae
Acrotriche serrulata (honey-pots) Epacridaceae
Nicotiana alata (wild tobacco) Solanaceae
Rhododendron laetum (rhododendron) Ericaceae

Plate 3   Floral Structure
Papaver nudicaule (Iceland poppy) Papaveraceae
Leucojum aestivum (snowflake) Amaryllidaceae
Scaevola pallida (coast fan-flower) Goodeniaceae
Leptospermum scoparium 'Lambethii' (manuka) Myrtaceae

Plate 4   Unusual forms
Amyema pendulum (drooping mistletoe) Loranthaceae
Allocasuarina verticillata (drooping she-oak) Casuarinaceae
Drosera glanduligera (scarlet sundew) Droseraceae
Cassytha melantha (coarse dodder-laurel) Lauraceae

Plate 5   Family Chenopodiaceae (saltbushes and samphires)
Atriplex cinerea (coast saltbush)
Suaeda australis (austral seablite)
Maireana triptera (three-wing bluebush)
Halosarcia pluriflora (samphire)

Plates

                Enchylaena tomentosa (ruby saltbush)
                Maireana sedifolia (pearl bluebush)
                Tecticornia verrucosa (tecticornia)

Plate 6   Various families
                Euphorbia wulfenii (spurge) Euphorbiaceae
                Zantedeschia aethiopica (arum lily) Araceae
                Brunonia australis (blue pincushion) Brunoniaceae
                Rhagodia candolleana (seaberry saltbush) Chenopodiaceae
                Rhagodia parabolica (fragrant saltbush) Chenopodiaceae
                Melaleuca armillaris (giant honey-myrtle) Myrtaceae

Plate 7   Legumes
                Acacia genistifolia (spreading wattle) Mimosaceae
                Acacia myrtifolia (myrtle wattle) Mimosaceae
                Acacia longifolia var. *sophorae* (coast wattle) Mimosaceae
                Trifolium repens (white clover) Fabaceae
                Dillwynia sericea (showy parrot-pea) Fabaceae

Plate 8   Family Solanaceae (nightshades)
                Cestrum parqui (green poison-berry)
                Solanum nigrum (black nightshade)
                Solanum laciniatum (large kangaroo apple)
                Solanum ellipticum (velvet potato bush, potato weed)
                Lycopersicon 'Burnley Sure Crop' (tomato)
                Salpichroa origanifolia (pampas lily of the valley)
                Brugmansia x *candida* (angel's trumpet)
                Nicotiana suaveolens (austral tobacco)

Plate 9   Various families including Myoporaceae (emu-bushes and myoporums)
                Myoporum sp. (sticky boobialla) Myoporaceae
                Myoporum insulare (common boobialla) Myoporaceae
                Eremophila glabra (common emu-bush, tar bush) Myoporaceae
                Eremophila maculata (spotted emu-bush) Myoporaceae
                Eremophila ovata (emu-bush) Myoporaceae
                Eremophila platycalyx (emu-bush) Myoporaceae
                Hibiscus rosa-sinensis (rose of China, hibiscus) Malvaceae
                Stylidium graminifolium (grass triggerplant) Stylidiaceae

Plate 10   Family Asteraceae (daisies)
                Galinsoga parviflora (gallant soldier)
                Chrysanthemoides monilifera (boneseed)
                Senecio elegans (purple groundsel)
                Calendula officinalis (garden marigold)
                Podolepis jaceoides (showy podolepis)

Sonchus oleraceus (milk thistle)
Hypochoeris radicata (flatweed)
Pycnosorus chrysanthes (golden billy-buttons)
Cynara scolymus (artichoke)

Plate 11   **Monocotyledons, Part 1**
Avena sterilis (sterile oat) Poaceae
Alopecurus pratensis (meadow fox-tail) Poaceae
Thysanotus tuberosus (common fringe-lily) Liliaceae
Wolffia australiana (wolffia) Lemnaceae
Caleana minor (small duck-orchid) Orchidaceae
Prasophyllum spicatum (leek-orchid) Orchidaceae
Thelymitra aristata (sun-orchid) Orchidaceae

Plate 12   **Monocotyledons, Part 2**
Lepidosperma gladiatum (coast saw-sedge) Cyperaceae
Ficinia nodosa (knobby club-rush) Cyperaceae
Juncus sarophorus (broom rush) Juncaceae
Cyperus eragrostis (umbrella sedge, drain flat-sedge) Cyperaceae
Xanthorrhoea australis (austral grass tree) Xanthorrhoeaceae

Acknowledgements

We are grateful to the many friends and colleagues who have assisted with the preparation of this book. We wish to thank Dr J. W. Anderson, Mr K. L. Blazé, Mrs J. J. Calder, Professor T. C. Chambers, Dr R. M. Foster, and Dr K. R. Thiele for their helpful comments on the manuscript, Mrs B. Joyce, Miss M. Carruthers, Miss S. Heard and Mrs M. Regan for their assistance with typing and proof reading, and Mr C. O'Brien and Mr T. Phillips for help with photography. We thank Mr A. W. Beudel for the photographs used in Plates 5a, 5e and 8i, Mrs C. Huggins for Plate 4f, and Mr T. Phillips for Plates 1f, 2b, 2d–e, 2g, and 4c. Finally we thank Professor T. C. Chambers and Dr P. M. Attiwill, chairmen of The University of Melbourne School of Botany, for their support during the course of this project.

For their help in the preparation of the second edition we thank Mr A. W. Beudel for photographs used in Plates 9h, 11g and 11h; Mr J. A. Jeanes for the photograph used in Plate 11i and for assistance with orchid nomenclature; Dr B. T. O. Lee for photographs in Plates 5f, 8c (left and right), 9e and 9f; Mr C. O'Brien for assistance with photography for Plate 11f, and the National Herbarium of Victoria, Royal Botanic Gardens, Melbourne, for access to facilities.

Introduction

For the amateur, whether gardener, bushwalker or naturalist, and even for the professional scientist who works in other fields of biology, there has been no readily available source of information on how to go about identifying flowering plants. There are many books of excellent photographs but if the plant of interest is not included, what do you do next? This introduction to the subject describes the structure of flowers and the process of identification, and we hope it will make the naming of flowering plants a little easier. It is intended to be used in conjunction with other published work such as field guides and Floras (a Flora is a book about the plants of an area usually with descriptions and keys for identification). The botanical language is often seen as a stumbling block and although we have explained many terms within the text, the glossary included covers most of the terminology used in current state and regional Floras.

The principles of plant identification are universal so the basic information is applicable anywhere. However, most of the illustrated examples occur naturally, or in cultivation, in south-eastern Australia.

The botanical names used for native plants are those accepted in the recently published *Flora of Victoria* and *Flora of New South Wales* (bibliography nos 219 and 104) but for one or two more recent changes. For the names of introduced plants grown in gardens in south-eastern Australia we have followed the *Horticultural Flora of South-Eastern Australia* (bibliography no. 209) or been guided by the staff of the National Herbarium of Victoria.

In this updated edition we have made a number of changes and added some new material. This has partly been in response to comments from reviewers and teaching staff who have provided useful feedback, and partly because of the need to keep abreast of change—with respect to plant names, to the large increase in identification literature and to the advent of computers.

The most significant changes and additions include:
- updating nomenclature to conform to that used in the recent Floras of Victoria and New South Wales

Introduction

- enlarging the treatment of the family Chenopodiaceae and adding treatments of the families Solanaceae and Myoporaceae with text and photographs
- enlarging the bibliography to include most of the large number of relevant books published in the last fifteen years
- adding notes on the relevance of computers to plant identification and the use of the internet in accessing information
- adding several general paragraphs on plant classification to Chapter 6 and on eucalypt classification to Chapter 8
- for selected families or species adding routes through the keys in several recent texts.

Biographical Note

After working in a variety of technical jobs, including with the Forests Commission of Victoria and the Botany Department of the University of Melbourne, Ian Clarke spent 14 years with the Botany Department of the University of Melbourne. He has most recently been employed in the Botanical Information and Identification Service at the National Herbarium of Victoria at the Royal Botanic Gardens, Melbourne. Between 1975 and 1990 he was associated with the Council of Adult Education in Melbourne as a tutor in flowering plant identification.

Helen Lee, Honorary Research Fellow in the Botany Department, La Trobe University, holds a Master of Science degree from the University of Melbourne. After lecturing in botany for two years at Queen's University, Belfast, she was for many years a part-time tutor in the botany departments of Melbourne and La Trobe Universities, and a tutor with the Melbourne Council of Adult Education. Helen's research interests include the morphological characteristics of plants that resprout after fire, and arid zone ecology. She is a member of the Royal Society of Victoria.

Their collaboration on this book brings together over thirty years of involvement in tertiary and adult education. Although they worked together throughout, Helen Lee is largely responsible for the text and Ian Clarke for the figures.

Symbols and Abbreviations

ACT	Australian Capital Territory	>	greater than
NSW	New South Wales	<	less than
NT	Northern Territory	+/-	present or absent
Qld	Queensland	∞	numerous
SA	South Australia	±	more or less, approximately
Tas.	Tasmania	♂	male
Vic.	Victoria	♀	female
WA	Western Australia	⚥	bisexual

cf.	compare	sp.	species (singular)
cv.	cultivar	spp.	species (plural)
plac.	placentation	ssp.	subspecies (also subsp.)
q.v.	which see (quod vide)	syn.	synonym
		var.	variety

L.S.	longitudinal section	T.S.	transverse section

mm	millimetre
cm	centimetre
m	metre

Floral formula symbols (used in captions) are discussed at the end of Chapter 2.

1

Getting Started

Background

The earliest known fossil plants, which resemble present-day mosses or liverworts, are over 400 million years old. In the Carboniferous period (360–286 million years ago) much of the land was low and swampy and supported vast forests of ferns and primitive seed-bearing plants. The preserved and compressed remains of many of these forests form the great coal deposits of the northern hemisphere; the brown coal deposits in Victoria are much younger and contain remnants of conifers and early flowering plants. About 275 million years ago the ancestors of present-day conifers (pines, firs, spruces etc.) appeared, and the flowering plants as we know them became dominant 100 million years ago. In contrast, the genus *Homo*, to which we belong, appeared about two million years ago.

The term botany is derived from a Greek word meaning 'plant' and is the branch of science that investigates plants. However, plants were important to the economy of humankind long before any scientific studies took place. People had to know which were good to eat, which were harmful, had medicinal value or could be used for making shelters, baskets and clothing. Some of the earliest non-religious books were the herbals which contained descriptions of plants, with their medicinal properties and recipes for herbal remedies. Plants containing hallucinogenic compounds are part of the folklore in many lands, having a particular place in religious ceremonies. One important ingredient of the witches' brews of medieval Europe was deadly nightshade, a relative of the potato and tomato. The use of cannabis is well documented; in Egypt, for example, it has been in use for over 4000 years.

Taxonomy is one of many branches of modern botany and deals with classification, that is, grouping and naming. But why are names so important? It is certainly easier to talk about a plant, or buy one for the garden, if you know what it is called. All plants have a scientific name that applies to one sort of plant only and has international recognition. Many amateur naturalists shy away from using them but they are useful in recognising relationships as well as being necessary for accurate identification.

Some essential information

In order to formally identify a plant, one needs to understand:
- the structure of the flower
- the botanical terms used in descriptions
- how to use reference books and, in particular, botanical keys
- something about plant relationships (the classification system)

These topics are discussed more fully in chapters 2 to 7; the following paragraphs introduce a few terms and concepts that form necessary background.

The system of naming plants now in use was established in 1753 by a Swedish naturalist Carolus Linnaeus. When Linnaeus put forward his scheme in the book *Species Plantarum*, all organisms were classified into one of two kingdoms: plants or animals. The criteria for their separation were that plants were immobile and neither ate nor breathed, whereas animals moved, breathed and ate. As new organisms were discovered they were assigned to one of these two kingdoms, but by the early twentieth century it had become obvious that this traditional division could not be sustained. Scientists have not yet reached agreement on how many kingdoms of organisms should be recognised. For introductory purposes, a simplified division of the plant kingdom can be made into algae (including seaweeds and freshwater forms), mosses and other moss-like plants, ferns and their allies, conifers (pines, cypresses, firs and their relations), and flowering plants. The fungi, including moulds, mushrooms and toadstools, are no longer considered to be part of the plant kingdom.

Many flowers are large, brightly coloured and scented, while others may be less conspicuous, but the plants that bear them all make up the group 'flowering plants'. Technical names proposed for this group include Angiospermae, Magnoliophyta and Anthophyta. There are two major subgroups of flowering plants, commonly known as the dicotyledons and monocotyledons: these terms are often contracted to dicots and monocots and that usage will be followed in this book. These two subgroups are further divided into orders, families, genera (singular—genus) and species. A species is a kind of plant, possessing a unique set of characteristics, and similar species are grouped into genera. A species has a two-word Latin name, e.g., *Grevillea robusta*; the first word is the name of the genus (*Grevillea*) and the second is the specific epithet (*robusta*). The common name of this plant is silky oak. Classification and nomenclature are dealt with more fully in Chapter 6.

Why is there such emphasis on floral structure? Linnaeus based his system of classification on the characteristics of the sexual reproductive organs and these are found in the flower. Knowledge of floral structure is the key to identification because even today it still forms the basis for plant classification, despite much additional information being available, for example, about anatomical, chemical and genetic characteristics. In addition, the structure of the flowers remains fairly constant, even if a species grows in a range of habitats which may result in considerable variation in the general appearance of the whole plant.

An appreciation of the structure of flowers and familiarity with the language will help you to see that plants are not isolated organisms, but form natural groups that are related to each other. Some groups are easy to spot: the orchids, wattles, gum trees and bottlebrushes, for example. However, the close relationships of genera in a family are not always immediately obvious, for example, the grouping of gum trees and tea-trees, or of proteas and grevilleas. We hope the use of this book will enable you to appreciate these relationships and increase your awareness of the amazing diversity and beauty of the world of flowering plants.

It is necessary to stress that this book does not present all there is to know about floral structure. It deals with the most common structural patterns and their variations, but some variations and exceptions are not included. The examples are drawn from plants common in gardens and the bush of south-eastern Australia, thus, many tropical families have not been mentioned. *However, the principles of identification are not geographically confined and the basic information is applicable in any region.*

Guidelines and conventions for pronunciation of botanical names can be found in *Australian Plant Genera* by Baines, *Plant Names* by Lumley and Spencer, *Western Australian Plant Names and their Meanings* by Sharr, and *The Names of Acacias of New South Wales* by Hall and Johnson (bibliography nos 15, 150, 204 and 103). Actual practice varies, particularly from country to country. Sometimes it is the loudest voice that holds sway. The standard ending for family names (–aceae) is usually pronounced, in Australia, as '*ay-see*'. For example the pea family Fabaceae = *Fab-ay-see*, and the eucalypt family Myrtaceae = *Mer-tay-see*. Sometimes a slight extra emphasis is placed on the last syllable, effectively pronouncing both the 'e' and the 'ae'. The standard ending for the names of orders (–ales) is usually pronounced '*ay-lees*'. Thus for the legumes, the order Fabales = *Fab-ay-lees*, and the order Myrtales (which includes the eucalypts) = *Mer-tay-lees*.

Using this book

One way of finding out about floral structure and coming to grips with the terminology is to work through the book from the beginning, referring to living plants as often as possible. Another approach is to compare the labelled illustration of a fuchsia or a grevillea, for example, with flowers from the garden. It is usually not essential to have the exact plant illustrated; often a related species will be sufficiently similar. The illustrations are specifically intended as a visual aid to the type of flower structure commonly seen in the various families. In a number of cases, two or more species in a genus have been drawn, allowing an appreciation of the close similarity in structure at this level. (See Figs 31–3, 41–4, 48–52, 80–4, 90–1, 130–1.) Unfamiliar terms can be checked in the glossary, which will often refer to an illustrated example in the text. Many illustrations are included in order to show the application of terms in a wide range of examples. Throughout the text, important terms appear in bold type; these are commonly encountered in books dealing with plant identification.

Notes on the illustrations

Figures 10 to 14 and the illustrations in Chapter 8 are representations of actual specimens, and are positioned to display as many of the relevant features as possible. To make comparisons easier, all flowers are drawn upright with the ovary at the bottom. This is sometimes not how they would naturally appear on the plant and some mental adjustment may be necessary. While the drawings are as accurate as possible, some natural variation must be expected and is part of the experience of getting to know plants. Slight exaggeration has been necessary in a few cases to improve clarity, such as enlarging the gap between closely adjacent parts.

The transverse section (T.S.) of the ovary in each case has been drawn to illustrate the placentation. It may not be a true 'section', as some depth has occasionally been added to make parts more obvious.

In sections and half flowers (which are exact halves), the cut surfaces are unshaded, and the adjacent surfaces have usually been shaded to try to make this clear. Shading within the ovary represents the loculus or cavity that contains the ovules. Figure 25 shows the way in which the flowers were cut to give the views drawn in most of the illustrations.

Labelling has occasionally been omitted from parts of drawings that were difficult to pinpoint accurately or clearly. In some cases, a composite structure, for example, the ovary, is difficult to label with a single arrow. Reference to the diagrams in Chapter 2 should clarify this.

The captions for figures 10–14 and 27–131 begin with the botanical name, *in italics*, of the species illustrated, followed by the common name (in parentheses), and the name of the family to which the species belongs. Usually a floral formula is included which summarises the basic floral structure shown in the drawing. These formulae are explained at the end of Chapter 2. The short description of the plant together with the illustration should provide enough information to enable the species to be 'identified' for practice in working with keys. This procedure is explained in Chapter 7.

Included in the captions are the magnifications of each part of the figures. Measurements made on the drawings, when divided by the magnification, will give the real dimensions. This will sometimes be very small—the ovary in *Leucopogon virgatus* (Fig. 91), for example, is about half a millimetre across.

The captions for the plates follow a similar pattern except that magnifications have been omitted—measurements given in the description will give an idea of size. Floral formulae are not included if details of the flower are not visible.

2

The Structure of Flowers

There are many ways in which the various flower parts can be arranged to make up the structure we recognise as a 'flower'. Consequently, it is difficult to reduce this floral variation to one or a few basic flower 'types'.

This chapter sets out to describe the structure of a simple flower, and then discusses some of the variation commonly encountered. Chapter 8 includes some more complex examples.

We have described floral structure under the following headings:

Structure of a basic flower
    The perianth
    The reproductive organs
Variation in floral structure
    Arrangement of parts
    The perianth
        The calyx
        The corolla
            Symmetry
            Aestivation
    The reproductive organs
        The androecium
        The gynoecium
            Placentation
            The style and stigma
            Discs
    Relationships of parts in the flower
    The floral tube
    The number of parts per whorl
    Unisexual flowers
    The floral formula

Structure of a basic flower

A basic flower (Fig. 1; Pl. 1a) has four series of parts arranged in concentric **whorls** (or rings) on the **receptacle**, which is the name given to the expanded end of the **pedicel** (flower stalk). The two outer whorls are together known as the **perianth**, and are not directly involved in reproduction. The male and female reproductive structures are located in the inner whorls.

The perianth

The outer whorl, known as the **calyx**, is composed of two or more parts called **sepals**, which are often green in colour and enclose the rest of the flower in the bud stage. Inside the calyx is the **corolla**, made up of **petals**, which are usually white or brightly coloured. It is usual for the sepals and petals to be equal in number.

The reproductive organs

A whorl of **stamens**, called the **androecium**, lies inside the corolla. Each stamen has a slender **filament** (stalk) and, at the top, an **anther** in which the **pollen** is produced. The pollen grains carry the male reproductive units.

In the centre of the flower is the **gynoecium**, made up of **carpels**. Each carpel has three parts: an expanded basal part called the **ovary**, in which the **ovules** are produced; a central stalk-like section called the **style**; and a terminal **stigma**. The ovules contain the female egg-cells and the stigma surface is specially adapted to receive compatible pollen. The ovules, after fertilisation and further development, become seeds and the mature carpels with the enclosed seeds develop into the fruit.

Figure 1b illustrates an expanded flower with the various whorls separated from each other. This places the receptacle at the bottom of the diagram, with the other parts laid out in sequence moving from the outside to the centre of the flower. Figure 1a shows the flower from above and illustrates the way in which the parts in adjacent whorls usually alternate with one another. The same flower from the side is illustrated in Figure 1c.

Variation in floral structure

Arrangement of parts

Although the majority of species have their flower parts arranged in whorls, others have numerous parts arranged spirally on a more or less club-shaped receptacle, for example, in the genus *Magnolia* (Fig. 10).

Name that Flower

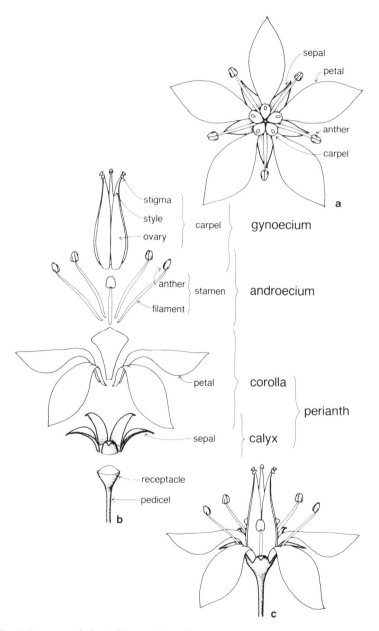

Fig. 1 Structure of a basic flower: **a** from above; **b** expanded flower, showing the series of parts; **c** side view

The perianth

THE CALYX

The sepals may be separate from one another as in Figure 1 or wholly or partly united (Figs 11, 74; Pls 2c, 8g). Sometimes an extra ring of sepal-like segments is attached outside the true calyx. This is known as an **epicalyx** and is typical of the genus *Hibiscus* (Pl. 9g) and of other members of the family Malvaceae (Fig. 11). Sometimes the sepals fall as the flower opens, as, for example, in many species of *Papaver* (poppy, Pl. 3a); in *Pittosporum undulatum* (sweet pittosporum, Fig. 44) the sepals may be shed before the petals drop. In other species the calyx persists when the flower fades and becomes a conspicuous feature of the fruit, for example, *Physalis* spp. (winter cherry, Cape gooseberry, Pl. 2c).

THE COROLLA

The petals may also be separate from one another or united. When united, the extent of the fusion and the shape and disposition of any free parts give rise to a variety of corolla types, some of which are illustrated in Figure 2 and Plates 2c—e, 3c, 8 and 9. The **corolla tube** is the term used to describe the usually tubular basal section, the **throat** is the top of the tube and the **limb** is the expanded part of the corolla above the throat and the tube (Fig. 2j). The term limb is also applied to the expanded upper section of a free petal that has a narrow basal part called a **claw** (Fig. 2i). Petals are considered united even if the degree of union is small, or is incomplete. This union may take the form of a very short tube, as in the genus *Sprengelia* (Fig. 95), or a tube in which the petals are free at the base, as in some species of *Correa* and *Stackhousia* (Figs 2k, 71). The corolla tube is then said to be **clawed**. In the family Goodeniaceae the petals are united but the corolla is split open down one side (Fig. 99; Pl. 3c).

Appendages on the petals occur in some groups, for example, the **corona** or trumpet of the genus *Narcissus* (daffodils and jonquils) and the **spur**, a tubular projection from a petal seen in *Velleia paradoxa* (spur velleia) and *Linaria* (toadflax).

The colour of the corolla may not be important in the identification of plants; for example, in *Epacris impressa* (common heath, Pl. 2d) it may be white, pink or red.

Sometimes the perianth segments, although in two whorls, are alike in size, colour and texture and then it is usual to retain the collective term perianth and to refer to the units as **perianth parts** or **tepals** (Figs 119, 121, 123). Such flowers are typical of the lily family, Liliaceae.

In some groups of plants there is only one whorl of perianth parts. The term perianth may be used to refer to the single whorl even if it is known to be a calyx or corolla. A single perianth whorl occurs in *Grevillea* (Fig. 33; Pl. 2b), *Clematis*

Name that Flower

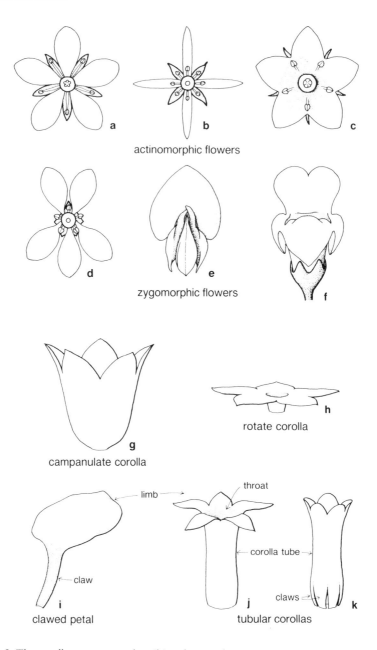

Fig. 2 The corolla: some terms describing shape and symmetry

(Fig. 38), and the widespread weed *Foeniculum vulgare* (fennel, Fig. 12). Sometimes all the perianth parts are absent or are reduced in size, often becoming brown and scale-like. In *Allocasuarina* (she-oak, Pl. 4c–d) the perianth is absent in the female flowers and reduced to fine brown scales in the male flowers.

Symmetry

As you look at a flower from above, the sepals and petals may be arranged on the receptacle in a symmetrical manner and the flower can be divided into equal parts by cutting through its centre in more than one plane (Figs 2a–c; Pl. 1a–c). Such a flower is said to be **actinomorphic** or **regular**. Alternatively, a flower may be asymmetrical and can be divided into equal parts by cutting in one plane only. This flower is said to be **zygomorphic** or **irregular** (Figs 2d–f; Pl. 2b, 3c). Some zygomorphic corollas are clearly **bilabiate** or two-lipped, with two of the petals grouped on one side of the flower and the other two or three on the other side (Fig. 2f; Pl. 9f). Members of the families Scrophulariaceae (snapdragons) and Lamiaceae (mints) typically have bilabiate corollas and, in the latter family, the calyx may also be bilabiate.

Aestivation

This is the term applied to the arrangement of the floral parts, usually the petals, in the bud. If the petals come together edge to edge they are said to be **valvate**, but when they overlap they are described as **imbricate**. Various other terms define more specialised arrangements, but these are not commonly encountered.

The reproductive organs

THE ANDROECIUM

Immediately inside the corolla are one or more stamens collectively called the **androecium** (shown in Fig. 1), a word derived from the Greek and meaning 'male house'. When numerous, the stamens may be in more than one whorl. Each stamen has a **filament** or stalk supporting an **anther**. The anther usually has two lobes, often lying parallel to one another, which are united by a tissue called the **connective**, effectively an extension of the filament (Fig. 3a). Each anther lobe contains two sacs in which the **pollen** is produced.

The filament may be attached at the base of the anther lobes—when it is said to be **basifixed** (Fig. 3b; Pl. 1d)—or at the back, **dorsifixed** (Fig. 3c; Pl. 1c). When the point of attachment is small, the anther may move freely with respect to the filament and is said to be **versatile**. If the filament is joined to the apex of the anther the lobes are described as **pendulous** (Fig. 3d), and if no longer parallel to one another are said to be **divergent**. Some stamens have extra appendages, for example, a projection of the connective above the anther lobes (Figs 3i, 64) or an extension of sterile tissue at the base of the lobes (Figs 3h, 118).

Name that Flower

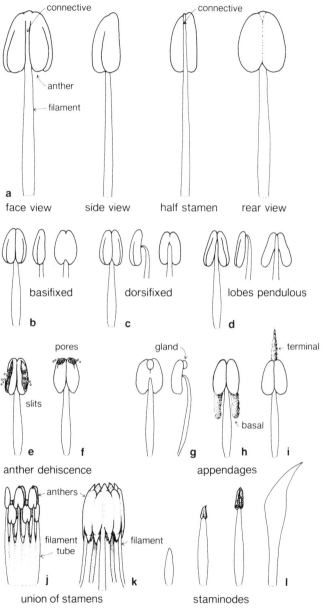

Fig. 3 The androecium—stamens: **a** stamen structure; **b–d** anther attachment; **e–f** anther dehiscence (see also Fig. 40); **g–i** appendages (see also Figs 64, 66, 118); **j** union of filaments (see also Fig. 59); **k** union of anthers (see Fig. 102); **l** staminodes

The Stucture of Flowers

When the pollen is mature, the wall of each anther lobe develops a longitudinal slit (Fig. 3e) or a terminal pore (Fig. 3f; Pl. 10c), through which the pollen is shed. The splitting of the anther, and the consequent shedding of pollen, is known as **dehiscence**. Each mature pollen grain contains a generative cell that, when conditions are favourable, eventually produces two male sperm cells.

Stamens are usually separate from one another, but may be united by their filaments or anthers. For example, in *Hibiscus* (Pl. 9g) and in *Ulex europaeus* (gorse or furze) the filaments are united for most of their length to form a tube, but the anthers are free (Figs 3j, 11, 59). In the genus *Melaleuca* the stamens are grouped in bundles due to the fusion of the filament bases (Fig. 87; Pl. 6f) and in members of the daisy family, Asteraceae, the anthers are united to form a tube around the style but the filaments remain free (Figs 3k, 102d).

The stamens are not always borne on the receptacle but may be attached to the petals or tepals. Stamens are said to be **epipetalous** (epi = upon) if attached to the petals as in *Epacris impressa* (common heath, Fig. 89) and *Solanum* (Pl. 8c, f), and **epitepalous** if joined to the tepals as in *Agapanthus praecox* (Fig. 123). Sometimes the stamens are reduced to anthers only. In *Grevillea* (Fig. 33; Pl. 2a), the anthers join directly to the tepals, there being no discernible filaments.

Occasionally the anther is reduced to one lobe, for example, in some stamens in the flowers of *Conospermum* (smoke-bush, Fig. 30). In some plants one or more stamens may be sterile due to a failure to produce pollen, or the anthers may be deformed or absent. In the latter case the filament may be reduced in length or modified to become petal-like (Fig. 3l). Abortive stamens are called **staminodes.**

In a flower with the same number of sepals, petals and stamens, the stamens nearly always alternate with the petals and so are opposite the sepals (Fig. 1; Pl. 1a). An exception is found in some members of the family Rhamnaceae in which the petals are opposite and hooded over the small stamens, for example, *Spyridium parvifolium* (dusty miller), or *Cryptandra amara* (Fig. 72).

THE GYNOECIUM

The innermost whorl of flower parts is called the **gynoecium** (Fig. 1), a term also derived from Greek and meaning 'female house'. The gynoecium consists of one or more structures called **carpels**. An external view of a single carpel is shown in Figure 4a. The swollen section at the base is the **ovary**, the stalk at the top of the ovary is the **style**, and near the end of the style is the **stigma**, a special tissue to which pollen can adhere and on which compatible pollen will germinate. (See the section on pollination in Chapter 4.) Figure 4a also shows one carpel that has been cut vertically from top to bottom. This view is known as a longitudinal section or L.S. The cavity inside the ovary, called the **loculus**

Name that Flower

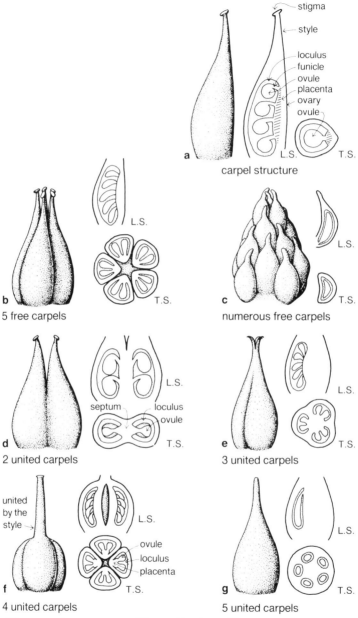

Fig. 4 The gynoecium—some possible combinations of free or united carpels: **a** structure of a single carpel; **b–c** free carpels; **d–g** united carpels

(plural = loculi) contains the **ovules**, which become the seeds following fertilisation and subsequent development. Each ovule is attached by a **funicle** (stalk) to the **placenta**, a tissue that lines part of the inside of the ovary wall.

A gynoecium of only one carpel is found in a number of large families including the Fabaceae (formerly Papilionaceae, the pea family), Mimosaceae (wattles) and Proteaceae, the family that includes *Grevillea* and *Banksia*.

Most flowers have two or more carpels, which are either free or wholly or partly united. A gynoecium of free carpels is said to be **apocarpous** (Figs 4b–c; Pl. 1a–b), and is typical of the families Magnoliaceae (Fig. 10), Ranunculaceae (buttercups, see Fig. 38, Pl. 1f), and Crassulaceae (stonecrops) which includes many succulent rockery plants, such as *Crassula* (Pl. 1a) and *Sedum* (Fig. 13).

A gynoecium of two or more carpels fused together into a single structure is said to be **syncarpous** (syn- = with, i.e. united). The degree of union can vary, and some of this variation is illustrated in Figure 4d–g and Plates 1c–e, 2f, 8c, f.

A gynoecium of two fused carpels is shown in Figure 4d. The ovary is completely divided into two **loculi** (cavities) and the wall between them is called the **septum** (plural = septa). In this case the septum bears the placentas. The styles and stigmas are free in this diagram indicating incomplete union, but often it is complete and there is then only one style and one stigma (Pl. 1d).

Sometimes a syncarpous gynoecium will contain only one loculus. In such cases the number of carpels involved will usually be indicated either by a divided style and/or stigma (Fig. 5e), or by the presence of more than one placenta within the ovary (Fig. 5g). An example of this type is seen in *Passiflora* (passion flower), which has a gynoecium composed of three united carpels with three styles and stigmas and a **unilocular** ovary. When a passionfruit is cut open and the pulp removed, the stalks to which the seeds were attached are seen in three groups on the inside of the fruit wall, indicating the location of the three placentas.

Partial fusion of the carpels also occurs in some genera of the Rutaceae, for example *Boronia* (Fig. 62), and in these flowers the carpels are free at the base but united above so that there is only one style and stigma (Fig. 4f).

In practice the term gynoecium is rarely used in literature directly concerned with plant identification, but is commonly replaced by the terms ovary and carpel. In botanical descriptions it is usual to read statements such as 'the ovary is trilocular', or 'the ovary is 1–4 locular'. The latter statement also illustrates the extent of the variation in carpel number in some groups. In some texts and in American literature, the term 'pistil' is often used, but it is sometimes confusing as it may refer to one of many free carpels in a gynoecium or to a syncarpous gynoecium.

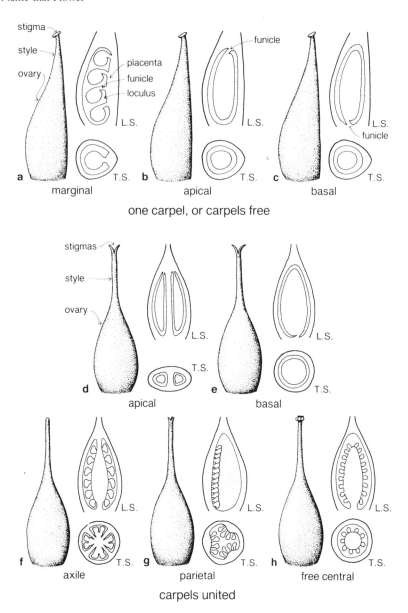

Fig. 5 Placentation—attachment of ovules in the ovary: placentation types are not restricted to the particular cases shown. For example, axile placentation (**f**) may occur with two or more united carpels.

How many carpels are in a flower?
Guidelines for finding the number of carpels when they are united
Cut a transverse section (T.S.) of the ovary and observe the number of loculi:

If the number of loculi > 1, then no. of carpels = no. of loculi

If only 1 loculus is present, then no. of carpels = no. of $\begin{cases} \text{stigmas} \\ \text{styles} \\ \text{placentas} \end{cases}$

Note Figs 4d–g and 5d–h and Pl. 2f. These guidelines will work in the majority of cases, but are not infallible. See also the sections on the families Lamiaceae (p. 185) and Goodeniaceae (p. 189).

Placentation

As mentioned earlier, the placenta is the tissue in the ovary to which the ovules are attached. The arrangement of the ovules (and hence the position of the placenta) within the ovary is termed the **placentation** and its determination is sometimes required when identifying a plant.

In a single carpel with numerous ovules the placentation is **marginal** (Fig. 5a). If there is only one ovule (or few) it may be attached at the top of the loculus, **apical** placentation (Fig. 5b) and then the ovule is said to be **pendulous**, or at the base of the loculus, **basal** placentation (Fig. 5c).

If the gynoecium is syncarpous (made up of two or more united carpels) and there are two or more loculi within the ovary, the placentas are usually situated on the central axis and the placentation is said to be **axile** (Figs 4d–f, 5f). However, if there is only one ovule (or few) in each loculus the placentation may be apical or basal (Figs 5d, 5e).

Other less common types of placentation occur when the ovary is unilocular and a number of united carpels are involved. For example, if within the loculus the placenta is raised on a central column the placentation is described as **free central** (Fig. 5h), and is typical of the family Primulaceae, for example *Anagallis* (pimpernel) and *Primula* (primrose). Again, a unilocular ovary may contain two or more placentas attached to its inside wall. This type of placentation is called **parietal** (Fig. 5g) and is found, for example, in *Passiflora* (passion flower). Another type of parietal placentation is seen in *Papaver* (poppy) in which plates of tissue carrying the placentas project inwards towards the centre of the loculus.

The style and stigma

The style is the tissue that links the top of the ovary with the stigma. It may be elongated (Figs 5d–h, 14), or relatively short (Figs 5a–c, 12), or occasionally absent (Pl. 3a). It is commonly inserted at the apex of the ovary (Fig. 6a), less frequently in a depression in the top of the ovary (Figs 6b, 89), or joined to the base of a deeply lobed ovary (Figs 6c, 98) when it is described as **gynobasic**. If the gynoecium is syncarpous, the style-end may be lobed (Figs 5g–h, 14) or the style may branch either close to the point of insertion on the ovary (Figs 12, 70) or nearer to the end (Figs 5d–e, Pls. 1e, 9g), and such branches are referred to as **style-arms** or **style-branches**. In the majority of cases the number of lobes or style-arms will equal the number of united carpels.

While the end of the style of a single carpel is usually undivided, it may be expanded in various ways; for example, in *Hakea* and *Grevillea* (family Proteaceae) the style-end is either conical or discoid (Figs 32–5; Pl. 2a, b).

The stigma is usually located on or near the end of the style or style-arms. The stigmatic surface may be covered with short glandular hairs or papillae (small rounded projections), which provide a receptive tissue on which compatible pollen may germinate. (See section on pollination, Chapter 4.)

Discs

Discs are not found in all flowers, but when present they are associated with the ovary, although not part of it. A disc at the base of an ovary is called **hypogynous** (hypo = below, gynous = female), while one across the top is an **epigynous** disc (epi = upon). The hypogynous disc may be a continuous ring of tissue (Fig. 7a), in the form of flat or rounded lobes (Fig. 7c) or it may appear as a swelling on one side of the ovary only (Fig. 7b). A disc is usually green, but may be coloured and, as it often produces **nectar**, it may also be referred to as a **nectary** or a nectary gland. Note the illustrations of *Boronia* (Fig. 62; Pl. 1c), *Zieria* (Fig. 68), *Grevillea* (Fig. 33; Pl. 2a), *Salpichroa* (Pl. 8f) and *Foeniculum* (Fig. 12), all of which include discs.

Relationships of parts in the flower

In most flowers, the parts of adjacent whorls alternate; thus the sepals alternate with the petals, petals with stamens, and stamens with carpels (Fig. 1). Another way of describing this arrangement is to say that the stamens are opposite the sepals and the carpels are opposite the petals. This typical pattern is an important clue to the interpretation of flowers with an unusual structure, for example, when one whorl of parts is absent.

The Stucture of Flowers

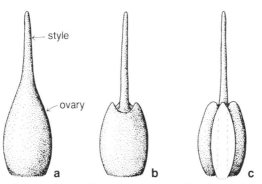

Fig. 6 Insertion of the style on the ovary: **a** terminal; **b** inserted in a depression; **c** inserted at the base of a deeply lobed ovary (gynobasic)

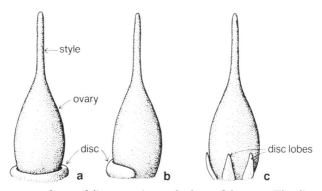

Fig. 7 Some common forms of disc occurring at the base of the ovary. The disc often produces nectar; it may also be referred to as a nectary (see also Figs 62, 32, 96).

The **relative** position of floral parts is one of the most important clues to the interpretation of floral structure.

When the flower parts are arranged on the receptacle so that the sepals occupy the lowest position, followed in order by the petals and stamens, with the carpels (whether free or united) at the top, the flower is said to be **hypogynous**, and the ovary is **superior** (Figs 1, 8a, 40; Pls. 1, 8c).

19

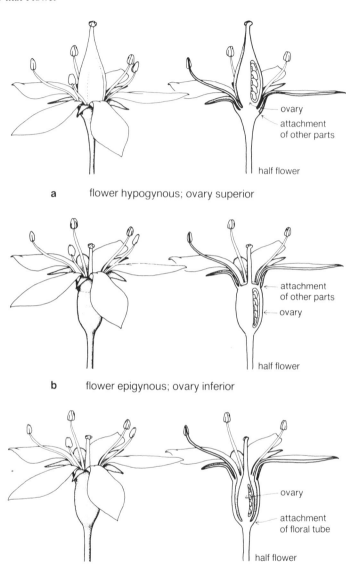

Fig. 8 Flower types and ovary position: three commonly recognized types of flower, and the corresponding ovary position with respect to the attachment of the perianth and stamens or the floral tube

Sometimes in a flower with a superior ovary, the receptacle appears to be cup-shaped so that the sepals, petals and stamens originate around the ovary. Such a flower is said to be **perigynous** (peri = around, Fig. 8c). The cup-like structure may be called the **floral tube**. Some genera of the family Rosaceae have perigynous flowers, including *Prunus* (cherry, Fig. 47) and *Rosa* (rose).

If the sepals, petals and stamens appear to come from the top of the ovary, the flower is said to be **epigynous** and the ovary is **inferior** (Figs 8b, 12; Pl. 3b–c). Occasionally, flowers appear to have an inferior ovary but the top of the ovary protrudes above the point where sepals, petals and stamens arise. Such an ovary is considered to be **semi-inferior**, and is seen, for example, in some species of *Pomaderris* (Fig. 73) and *Leptospermum* (Pl. 3d).

The ovary is superior (above) or inferior (below) with respect to the point of attachment of the other floral parts (perianth and stamens).

The floral tube

The floral tube may be defined as consisting of the fused bases of the perianth parts (both calyx and corolla if both are present) and the stamens (Fig. 9a–b). (Note that one of the perianth whorls may be absent, as in *Pimelea*—see Fig. 76.) It should be clearly distinguished from a corolla tube which involves the union of petals only, as shown in *Epacris* (Fig. 89; Pl. 2d).

The floral tube has been the subject of much debate, particularly with respect to its derivation, and has been given many names by various authors. The term 'floral tube' is not widely used in the literature. The most common alternative is **hypanthium**, and others used include **thalamus tube, receptacle tube** and **torus**. Sometimes the term **calyx tube** is used in this sense but is not strictly correct. Note that the term 'torus' is also synonymous with 'receptacle', implying the interpretation that the floral tube is an outgrowth from, or part of, the receptacle. A flower may thus be described as having a 'hollow cup-shaped torus'. This interpretation is used extensively in the *Flora of South Australia* by J. M. Black.

Sometimes the ovary is described as being 'adnate to the floral tube'. This implies an inferior, or at least semi-inferior, ovary. In this sense the floral tube may not be evident as a separate structure, as the tissues of the ovary wall and floral tube are integrally fused. In the example of *Fuchsia* (Fig. 14), the floral tube may be described as 'adnate to, and produced above the ovary'.

The presence of a floral tube should not be a problem to those identifying plants except that it may affect the decision about freedom or union of perianth parts. When a floral tube is present and the sepals and petals are distinctly different, and arise independently from the top of the tube, they are usually regarded as being free (Fig. 9a, *Leptospermum* Fig. 86; Pl. 3d).

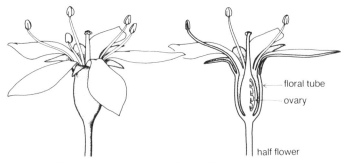

a floral tube present; sepals, petals and stamens free; ovary superior

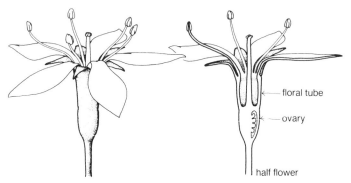

b floral tube present; sepals, petals and stamens free; ovary inferior

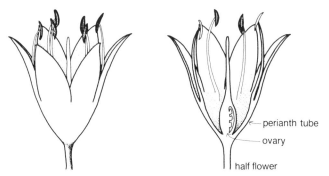

c floral tube not recognized— perianth tube present; tepals united, stamens epitepalous; ovary superior

Fig. 9 The floral tube

The Stucture of Flowers

When a floral tube is technically present but the perianth is not obviously differentiated into sepals and petals, the floral tube may be ignored and the tepals described as united.Then, the stamens can be regarded as joined to a perianth tube, and the term epipetalous stamens applies (Figs 9c, 123). In these cases the tube usually appears to be a continuation of the perianth.

Floral tubes are encountered in members of a number of families, such as:

Myrtaceae	(eucalypts, tea-trees, bottlebrushes—Fig. 78 etc.)
Onagraceae	(*Fuchsia* Fig. 14, evening primroses)
Rhamnaceae	(*Cryptandra* Fig. 72)
Rosaceae	(roses, *Acaena* Fig. 45, *Chaenomeles* Fig. 46, *Prunus* Fig. 47)
Thymelaeaceae	(*Pimelea* Fig. 76).

When a floral tube is present, that is, the bases of all perianth parts and stamens are united—
decide about the freedom or union of parts
- at the top of the tube if the sepals and petals are obviously different
- at the bottom of the tube if the sepals and petals are similar in size, colour and texture

Number of parts per whorl

The number of units in each whorl of the flower varies from species to species. The degree of variation is considerable, ranging from zero to numerous (the latter term usually is applied when the number of parts exceeds 10). The units commonly occur in multiples of the petal number, for example 5 sepals, 5 petals, 5 or 10 stamens and 5 carpels. In the dicots the basic number of units per whorl is commonly 4 or 5, and in the monocots the units are usually in multiples of three (for explanation of dicots and monocots, see Chapter 6). Reduction in the number of parts is common; for example, often there are only two carpels in a 4- or 5-partite flower, or fewer stamens than petals (Pl. 9a).

Unisexual flowers

Most flowers have both stamens and carpels and are therefore bisexual. However, some have no stamens and thus are unisexual and functionally female. Similarly, flowers without carpels are functionally male. Unisexual flowers may occur on the same plant, as in cucumbers, marrows and pumpkins and sometimes in *Allocasuarina* (she-oak). Such plants are said to be **monoecious**. When the male and female flowers are borne on separate individuals the species is said to be **dioecious**; examples include *Clematis* (Fig. 38), *Wurmbea dioica* (early nancy, Fig. 121), some species of *Atriplex* (Pl. 5a), and most species of *Allocasuarina* (Pl. 4c–d).

Name that Flower

The floral formula

A floral formula is a shorthand means of recording the basic structure of a flower. Standard symbols used for the floral organs are listed below. Formulae have been used in most captions; they have been expressed as simply as possible to indicate what is readily visible in the flower.

P   perianth   — tepals
K   calyx   — sepals
C   corolla   — petals
A   androecium — stamens
G   gynoecium — carpels
     G   ovary superior
    –G– ovary semi-inferior
     G̅   ovary inferior
∞   parts numerous (>10)
( ) parts of a whorl united
{ } used in this book to indicate parts of a whorl at first united and becoming free as the flower ages
⌒ links united parts of different whorls
+   indicates parts are in more than 1 whorl
,   separates numbers of parts within a whorl

EXAMPLES

Fig. 1

K5	C5	A5	G5̲
sepals 5	petals 5	stamens 5	carpels 5
free	free	free	free
			ovary superior

Sedum Fig. 13

K5	C5	A5+5	G5̲
sepals 5	petals 5	stamens in 2 whorls	carpels 5
free	free	5 epipetalous 5 free	free ovary superior

Epacris Fig. 89

K5	C(5)	A5	G(5̲)
sepals 5	petals 5	stamens 5	carpels 5
free	united	free epipetalous	united ovary superior

24

Foeniculum Fig. 12
K0	C5	A5	G($\overline{2}$)
sepals absent	petals 5	stamens 5	carpels 2
	free	free	united
			ovary inferior

Lathyrus Fig. 58
K(5)	C(2),3	A(9),1	G$\underline{1}$
sepals 5	petals 5	stamens 10	carpel 1
united	2 united	9 united	ovary superior
	3 free	1 free	

Ricinocarpos Fig. 70 (male flower)
K(5)	C5	A(∞)	G0
sepals 5	petals 5	stamens numerous	carpels absent
united	free	united	

Grevillea Fig. 33
	$\widehat{P(4)}$	A4	G$\underline{1}$
	tepals 4	stamens 4	carpel 1
	becoming free	free	ovary superior
		epitepalous	

In a few cases, we have used the form 'G(loculi)' to indicate the number of loculi visible in a transverse section of the ovary, instead of noting the number of carpels in the gynoecium. (See Figs 79, 97–9.) For further reading on the structure of flowers, see bibliography numbers 33, 143 and 181.

Name that Flower

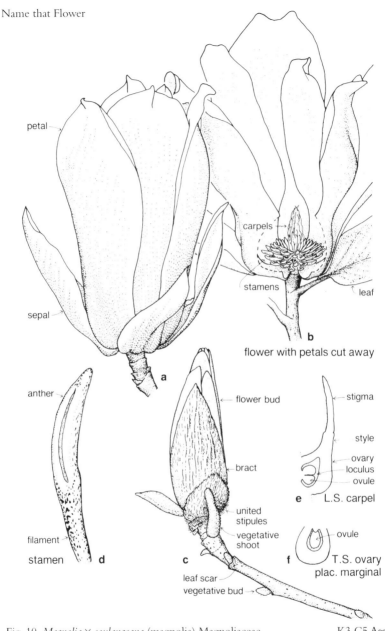

Fig. 10 *Magnolia* × *soulangeana* (magnolia) Magnoliaceae   K3 C5 A∞ G∞

Large shrub or small tree to 4 m or more tall. Leaves ovate-oblong, deciduous, with deciduous stipules. Flowers terminal. Petals variable in number, white inside, purplish-pink on the outside. A hybrid, originating in France. Widely cultivated. Flowering in spring before the new leaves. (a–c ×0.6, d–f ×6)

The Stucture of Flowers

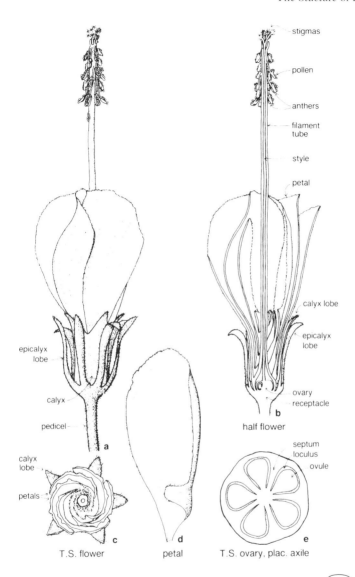

Fig. 11 *Malvaviscus arboreus* (wax mallow) Malvaceae K(5) C͡5 A(∞) G(5)

Spreading shrub to 2 m. Leaves broadly ovate, to 15 cm long, hairy, becoming 3-lobed at the apex. Flowers red, solitary in the axils. An epicalyx of bracts surrounds the tubular calyx. The petals are joined to the filament column which surrounds the style. Cultivated, origin Central to South America. Flowering in summer to autumn. The family includes *Hibiscus* (Pl. 9g), *Althea* (hollyhock) and weeds such as the mallows. (a–b ×2, c ×2.5, d ×2, e ×7)

Name that Flower

The Stucture of Flowers

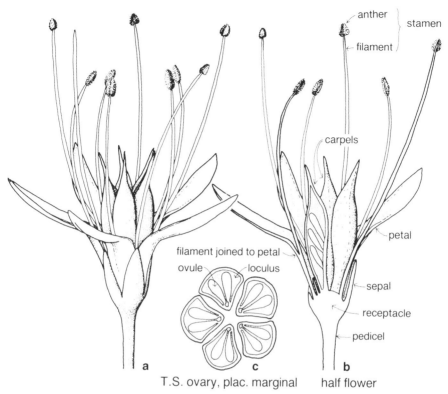

Fig. 13 *Sedum spectabile* (stonecrop) Crassulaceae   K5 C5͡ A5+5͡ G5
Succulent perennial herb less than 1 m high. Leaves opposite, obovate, about 6–8 cm long. Flowers pink, borne in dense corymbose cymes. The sepals are very slightly united but this is usually ignored. Stamens of one whorl are epipetalous. Carpels free. Cultivated, origin China and Korea. Flowering in autumn. The family includes many succulent rockery plants. See also Pl. 1a. (a–b ×7, c ×12)

Fig. 12 *Foeniculum vulgare* (fennel) Apiaceae (Umbelliferae)   K0 C5 A5 G($\overline{2}$)
Glabrous perennial herb to 2 m high. Leaves compound, dissected into fine segments. Flowers yellow, small, borne in compound umbels. The top of the ovary is covered by a yellow disc. Stamens mature and drop off in sequence, before the styles and stigmas develop. Fruit dry, indehiscent. Widespread weed. Origin southern Europe. Flowers in late spring to early autumn. The family is sometimes called Umbelliferae, and also includes carrot, parsely and celery. (a ×0.5, b–e ×12)

29

Name that Flower

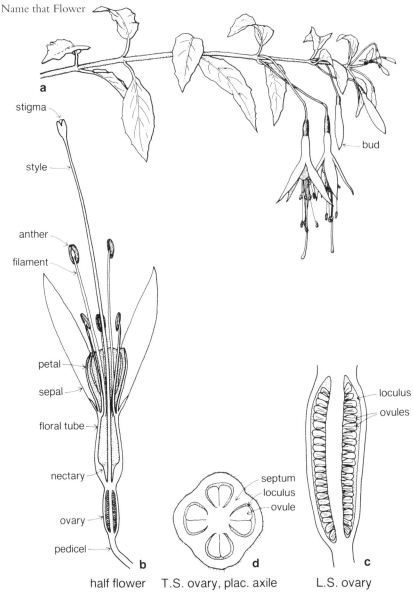

half flower   T.S. ovary, plac. axile   L.S. ovary

Fig. 14 *Fuchsia magellanica* (fuchsia) Onagraceae   K(4) C4 A4+4 G($\bar{4}$)
floral tube present

Erect shrub to 2 m, branchlets drooping. Leaves opposite, ovate-lanceolate, to 8 cm long, margins serrate. Flowers red and purple, in upper axils. Sepals united to a small degree. Nectary present at the base of the floral tube. Introduced from South America and commonly grown, sometimes escaping from cultivation, as in the Otway Ranges, Vic. Flowering summer and autumn. (a ×0.7, b ×2, c ×7, d ×12)

3

Inflorescences—the arrangement of flowers on the plant

An **inflorescence** is a flowering shoot that carries more than one flower. The following is an introduction to the basic terms used to describe the arrangement of flowers.

Solitary flowers

Flowers are said to be **solitary** when the plant bears only one, or when single flowers appear on the ends of lateral branches that are remote from one another. In practice, the inflorescence of a shrub with one flower in each leaf axil on many shoots is often described as 'flowers solitary in the axils of the leaves'.

Racemose inflorescences

Racemose inflorescences (Fig. 15) are characterised by a pattern of branching termed monopodial. The main axis continues to grow, producing lateral buds, which become flowers or shoots that repeat the same pattern. The youngest buds are nearest the apex. The racemose inflorescence types shown in Figure 15 are defined in the glossary.

Cymose inflorescences

In cymose inflorescences, the growth of the main axis is terminated by a flower. One or more lateral buds behind that flower then develop until their growth is in turn stopped by the production of flowers. This branching pattern is termed sympodial, and some examples are shown in Figure 16. In practice, flowers may simply be described as being in 'cymes'. This indicates an inflorescence of this general type without specifying a particular arrangement.

Name that Flower

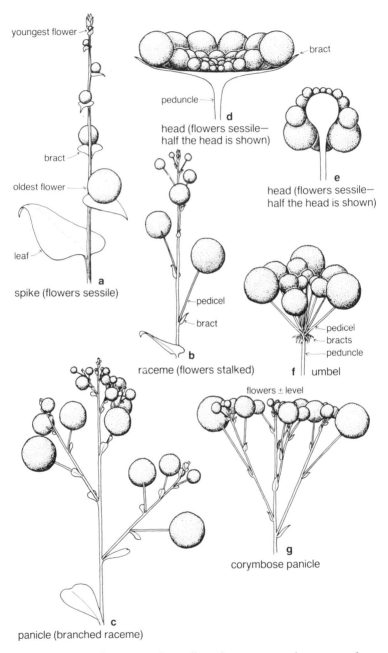

Fig. 15 Racemose inflorescences: the smallest spheres represent the youngest flowers

Inflorescences

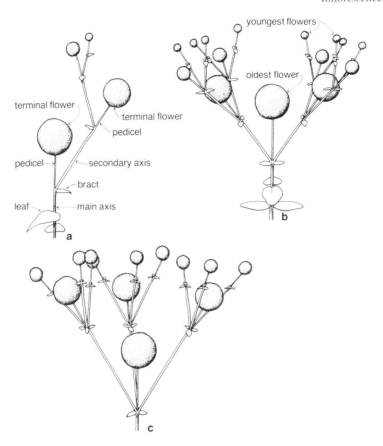

Fig. 16 Cymose inflorescences: **a** compound monochasium—in practice, the stem may not be so obviously zigzag (a simple monochasium is similar but consists of only the two lowest flowers); **b** compound dichasium (a simple dichasium consists of the first terminal flower and two lateral flowers only); **c** a cymose inflorescence as seen in *Anemone*

Cymose inflorescences are common in the family Caryophyllaceae, which includes *Silene* (catchfly), *Lychnis* (campion) and *Gypsophila* (baby's breath), as well as the chickweeds.

Name that Flower

Some inflorescence types, such as the head and umbel, can be either racemose or cymose. When cymose, the youngest flowers are towards the outside of the inflorescence, instead of towards the centre as shown in Figures 15d–f.

Sometimes inflorescences do not fall neatly into a particular category and terms may be used rather loosely. For example, the panicle, although defined as racemose, is often used for any multi-branched inflorescence as we have done in the following list. Alternatively, terms may be qualified, such as 'spike-like panicle' which is used where a branched inflorescence is dense and slender, the branches short and not readily seen. The term 'cluster' is often used in a general sense and does not imply any particular branching pattern.

Figure and plate numbers of illustrations of various inflorescence types are listed below.

Inflorescence type	Family, genus, etc.	Figure	Plate
Solitary	*Caladenia*	126	
	Chiloglottis	128	
	Epacris	88	2d
	Papaver		3a
Spike	*Acacia floribunda*	49	
	A. longifolia var. *sophorae*		7c
	Callistemon	77	
	Melaleuca		6f
	Stackhousia	71	
Raceme	*Drosera*		4e
	Grevillea	31, 33	
	Lathyrus	56	
	Senna	53	
	Stylidium	100	
	Thomasia	74	
Panicle	*Conospermum*	29	
	Zieria	67	
Umbel	*Agapanthus*	122	
	Eucalyptus	80, 83	
Compound umbel	*Foeniculum*	12	
Head	*Acacia*	48, 50–52	7a–b
	Asteraceae	101–10	10 (except i)
	Brunonia		6c
	Pimelea	76	
Compound head	*Calocephalus*	111	
	Pycnosorus		10i

4

Reproduction

Reproduction in the flowering plants may be either sexual or asexual. Sexual reproduction involves the fusion of male and female reproductive cells known as gametes. In asexual reproduction no fusion of gametes takes place and the reproductive entities are vegetative bodies such as corms, tubers, bulbs, cuttings etc.

Sexual reproduction may be dealt with in a number of stages:
- pollination—the transfer of pollen from anthers to stigma
- fertilisation—union of male and female gametes
- development of the seed
- development of the fruit
- germination of the seed.

Pollination

The transfer of pollen, which carries the male gamete, to the stigma takes place in several ways. Many plants are pollinated by wind including the conifers (pines and their allies), she-oaks and grasses, and many northern hemisphere trees such as birches, alders and oaks. These plants produce large quantities of pollen and release it into the air and it may float or be blown on to a compatible stigma. The stigmas of wind-pollinated plants are commonly large and feathery, which gives them a better chance of trapping pollen.

Insects are the most important pollinators of flowers. They visit to collect nectar or pollen or both, and at the same time incidentally transfer some pollen from one flower to another. Insects locate flowers by odour and then are influenced by colour and shape. It is known that bees preferentially visit yellow or blue flowers whereas moths, which emerge in the evening, are attracted to white or cream flowers, which are more readily seen at that time.

Birds, particularly honeyeaters, are important pollinators of flowers with tubular corollas, to which they are attracted by the presence of copious nectar. Pollen catches on the head feathers of the birds as they probe for nectar and then is carried to other flowers. Birds seem to be attracted to red flowers but they visit other colours if nectar is available. Small animals such as pygmy possums, glider possums and dibblers are also believed to pollinate certain species.

Insect- and bird-pollinated plants usually have large colourful flowers and produce nectar. In contrast, the flowers of wind-pollinated plants are commonly small, green or brown, often with no petals and lacking nectar.

Fertilisation

At maturity most pollen grains contain three nuclei, two of which are sperm, or male gametes, while the other, known as the tube nucleus, appears to be involved with the growth of the pollen-tube.

The mature ovule contains an embryo sac, and the most usual type has eight nuclei enclosed in seven cells. The egg-cell, or female gamete, flanked by two other cells is at one end of the embryo sac; in the centre are two polar nuclei and at the other end the three remaining cells.

When a pollen grain germinates on the stigma, the pollen-tube emerges and grows down through the style, through the carpel wall and into the ovule and embryo sac. The tube nucleus moves down with the advancing end of the pollen tube followed by the two sperm nuclei (Fig. 17). The sperm nuclei enter the embryo sac, and one fuses with the egg-cell to form the zygote, which is the

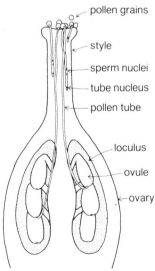

Fig. 17 Diagrammatic longitudinal section of two united carpels after pollination; the pollen tubes are seen growing down the style

general name given to a fertilised egg in all organisms. The second sperm nucleus fuses with the two polar nuclei to form the primary endosperm nucleus, which divides very rapidly to produce a tissue called the endosperm.

Germination of the pollen depends on a chemical interaction between exudates from the wall of the pollen grain and the surface of the stigma. A favourable reaction between the exudates enables the pollen to germinate and the pollen tube will grow down into the style. In such a case the pollen is said to be compatible. If the reaction results in the suppression of pollen germination, the pollen is incompatible. Many plants are self-incompatible, which means that their own pollen will not germinate on the stigmas of the same individual. The plants must then be cross-pollinated with pollen from another individual of the same species in order to be fertilised. Usually stigmas reject pollen from other species but when different species do fertilise one another they are said to have hybridised.

Development of the seed

After its formation, the zygote divides and develops into an embryo, this process, at least in part, absorbing nutrients from the endosperm. A mature embryo consists of an axis (the shoot–root system of the future) and either one or two cotyledons, sometimes called seed-leaves. Generally a dicot embryo will have two cotyledons and a monocot embryo only one. (See Chapter 6 for explanation of 'dicot', and 'monocot'.) In dicots the remaining food stored in the endosperm may be absorbed by the cotyledons, which become thick and fleshy as in beans, peas and wattles, or the endosperm may remain in the mature seed as in the castor oil plant and some members of the Chenopodiaceae (saltbush family). Monocot seeds all contain endosperm that is absorbed through the cotyledons during germination. Mature embryos are surrounded by a seed coat or **testa** (Fig. 18) but in many monocots, such as the grasses, the testa is fused with the fruit wall. The **hilum** (Fig. 18) is the scar on the seed marking the point where the funicle was attached.

Development of the fruit

Fruits either form from the gynoecium of a single flower or, less often, from an inflorescence. **Simple** fruits develop from a single carpel or a syncarpous gynoecium, and are dealt with in more detail below. A flower with free carpels gives rise to an **aggregate** fruit (Pl. 1f) and examples include the strawberry and raspberry. (See the section on the family Rosaceae, p. 103.) **Multiple** fruits are formed from an inflorescence; two examples are the pineapple and the fig. The pineapple consists of many fleshy units attached to a central axis, and the pattern on the tough outer skin shows the boundaries of the individual fruitlets.

In the fig, the flowers (and later the small fruitlets) are enclosed in a hollow inflorescence axis which becomes fleshy as it matures.

The fruit wall, or pericarp, generally develops from the carpel wall or, as in apples and rose hips, it may include the floral tube. The pericarp is sometimes clearly differentiated into three layers and then the outermost layer is the exocarp, the middle one the mesocarp and the inner the endocarp. The three layers are present in a cherry: the skin is the exocarp, the flesh the mesocarp and the stone that encloses the seed the endocarp. In fruits that are dry at maturity the layers are incompletely differentiated. Sometimes the perianth remains attached to the fruit and may enlarge as the fruit grows (Pls. 2c, 8d). In such cases, the terms 'fruiting perianth' or 'perianth persisting in fruit' are used.

At maturity, fruits may be fleshy, or hard and dry. Those with dry pericarps are either dehiscent (split open to release their seeds) or indehiscent. Sometimes a dry fruit produced by a multilocular ovary splits up to form several fruitlets, as in *Correa* and some other members of the Rutaceae.

In botany the term fruit includes many so-called vegetables such as peas, beans, cucumbers, capsicums and pumpkins. The section dealing with the Rosaceae describes some familiar fruits belonging to that family. The accompanying list of fruit types commonly encountered in literature dealing with plant identification includes some well-known examples. The fruit types are defined in the glossary.

    Fruit dry
        Fruit dehiscent
            Follicle *Macadamia*, *Banksia* (Fig. 27),
                      Grevillea (Fig. 31), *Hakea* (Fig. 34)
            Legume bean, lentil, pea, peanut
                      Senna (Fig. 53)
                      Lathyrus (Fig. 57)
            Capsule *Callistemon* (Fig. 77),
                      Eucalyptus (Fig. 80), tea-tree
        Fruit indehiscent
            Nut   acorn
            Cypsela individual fruitlets of the daisies
                      Tagetes (Fig. 102),
                      Senecio (Fig. 106), *Taraxacum* (Fig. 109)
            Achene individual fruitlets of the buttercup (Pl. 1f)
    Fruit fleshy
        Drupe cherry, date, plum
        Berry   egg plant, guava, kangaroo-apple (Pl. 8c)
                  passionfruit, tomato, *Rhagodia* (Pl. 6e)

In the natural environment fruits show a variety of adaptations that aid their dispersal, and hence the spread of their seeds. Birds commonly eat fleshy fruits and then often deposit the seed well away from the fruit source. The fruits of plants such as *Acaena* (sheep's burr, Fig. 45, bidgee-widgee) and *Medicago* (medic) have awns that become caught in the fur of animals, not to mention people's socks, and the fruits may be carried some distance from the parent plant. The feathery tufts on thistle and daisy fruits and the wings on fruits of elms, birches and ashes assist in their dispersal by wind. Some plants form pods that open explosively causing the seeds to be catapulted out. The seed of the coconut palm is well protected by a thick fibrous husk, and coconuts have drifted on ocean currents to many remote islands. Other plants that live close to beaches are known to spread in the same way.

Germination

If the seed of a pea or bean is soaked in water for a time the testa can be removed easily (Fig. 18). Inside are the two fleshy cotyledons and if these are opened out the embryo plant can be seen between them. The cotyledons are joined to the axis of the embryo at a point known as the cotyledonary node. The axis above the node is called the epicotyl, which bears the apical meristem and often the first pair of leaves as well (Fig. 18). Below the node the axis is called the hypocotyl and the embryonic root at its lower end is the radicle.

At germination the young root usually emerges first and develops sufficiently to anchor the young plant in the ground. In plants such as beans, wattles and eucalypts, the hypocotyl elongates and pushes the seed above the soil surface, whereupon the cotyledons emerge from the testa, spread out and become green; then the epicotyl elongates and the first two foliage leaves spread out and begin to grow. In plants such as the edible peas the epicotyl elongates first and the seed and cotyledons remain in the soil. During germination the plant absorbs nutrients from the cotyledons or endosperm until it can sustain itself by photosynthesis (see Chapter 5). Monocot seeds are very variable in structure but the general principles apply, and in all cases the single cotyledon absorbs nutrients from the endosperm to support the growth of the young plant.

Many seeds will not germinate immediately after their release from the fruit. They enter a period known as dormancy, which under natural conditions ensures that germination will not occur until the chance of survival for the seedling is greatest. A number of plants, particularly those in regions where the winters are very cold, produce seeds that must be held at a low temperature for some time before they will germinate. This process, known as stratification, can be simulated by storing the seeds in a refrigerator. In nature the low soil temperature ensures that germination will not occur until the conditions are such that the seedlings are likely to survive.

Some seeds, such as those of wattles, have very tough testas that must either gradually rot away or be cracked open by the heat of a fire. After fire, large numbers of seedlings will emerge in the open habitat, free of competition and, with extra nutrients available from the burnt vegetation, these seedlings have an excellent chance of survival. Many Australian plants produce seeds that can apparently remain viable in the soil for a long time and germinate after fire.

The testas of the seeds of some desert plants contain inhibitory chemicals, which must be leached away by the soil water before germination will take place. In another group of plants, the seeds need to pass through the intestine of a bird or animal before they will grow. Sometimes this process can be simulated by acid treatment but in other cases the effect of the animal is not understood.

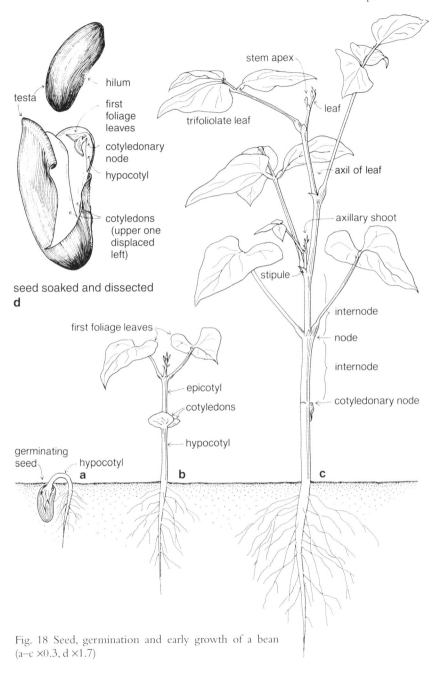

Fig. 18 Seed, germination and early growth of a bean (a–c ×0.3, d ×1.7)

5
Introduction to Plant Structure and Function

A typical plant and the arrangement of the various parts is illustrated in Figure 18. Most plants have a stem–root system called the axis. The stem supports the leaves, flowers and branches and the root anchors the plant in the soil.

Cells

All living organisms have cells as their structural units. Plant and animal cells are similar in many respects but differ in some fundamental ways. Plant cells have a strong wall made of cellulose, a complex carbohydrate of glucose molecules linked to form cellulose molecules which group into rigid fibrils. Within the cell wall a membrane encloses the cytoplasm, which is a complex assemblage of membranes, organelles and vacuoles (cavities) all suspended in a watery fluid. The vacuoles, which are enclosed by other membranes, contain various enzymes as well as waste products of the cell. In mature plant cells the vacuoles are usually consolidated, and one large vacuole occupies most of the cell; a thin layer of cytoplasm separates the vacuole from the cell membrane.

One type of organelle, found in plants and most algae, is called a chloroplast. The internal membranes of chloroplasts contain pigments, the most prominent being chlorophyll, which imparts a green colour to the chloroplasts and to plant cells exposed to light. The nucleus is the largest organelle within the cell and it contains a compound called deoxyribonucleic acid (DNA) in which is encoded all the hereditary information that controls cell structure and function. During cell division the DNA is associated with proteins to form the chromosomes, which transfer the DNA from one generation to the next.

Stems

Most stems may be classed as **herbaceous** or **woody**. Herbaceous stems are usually green and relatively soft. They are characteristic of annual plants such as peas and beans, which complete their life cycle in one season, and of those plants, including some sundews and orchids, that shoot each season from an underground resting organ. The stems of trees and shrubs, which are perennial plants, are commonly woody and are hard and tough by comparison with the herbaceous type.

Plants have a variety of growth forms. Many are erect and self-supporting, but scrambling or climbing plants need outside support in order to grow upwards. The stems of climbers such as *Wisteria* and *Comesperma volubile* (love-creeper), the petioles of *Clematis*, and the modified branches, known as tendrils, of *Vitus* (grape) twist around any convenient object, usually another plant. The stems of prostrate plants lie flat, and some put down new roots at intervals. Strawberries are said to be stoloniferous, as they form new rooted shoots at the ends of horizontal stems called **stolons** or runners.

During the course of evolution, stems of some species have been modified to perform a variety of functions. In some plants such as cacti, the stem is often flattened and green and the leaves are absent or rudimentary. Such a stem is called a **cladode**. Cacti are also succulent, containing water-storage tissues. The potato **tuber** is a fleshy underground stem containing stored food; on its surface are vegetative buds, each of which can produce a new shoot. Orchids and sundews are among many plants forming tubers that remain in the soil at the end of one growing season and give rise to a new plant in the next. **Corms** are condensed stems full of stored food and with surface buds, as in *Gladiolus*. In **bulbs** such as onions, fleshy leaf bases are borne on a short stem. Around the outside of the bulb, dry and often papery leaf bases form a protective layer. **Rhizomes** are underground stems that run more or less parallel to the soil surface and give rise to aerial shoots at intervals. Familiar rhizomatous plants include bracken fern and some irises, sedges and rushes.

Leaves

The points where leaves are attached to stems are called **nodes** and the section of stem between two nodes is termed an **internode** (Fig. 18). A leaf (Fig. 19) has two principal parts, the **lamina** or blade and the **petiole** or stalk. When there is no petiole the leaf is said to be sessile. The lamina is supported by a system of veins, and when a main vein or midrib with numerous branches forms a network the **venation** is said to be reticulate. This pattern is typical of the dicots. When several veins of approximately equal size run parallel to each other the venation is described as parallel and this arrangement is common in the monocots (Fig. 19 b–c).

A leaf with a single blade is described as **simple**, and those with more than one blade as **compound**, and the component leaflets are called pinnae. A compound leaf is **pinnate** when the leaflets arise from the rachis, a continuation of the petiole, and **palmate** when the leaflets diverge from the end of the petiole. Pinnae may be further divided into pinnules, and then the compound leaf is said to be **bipinnate**.

Leaf arrangement is described by terms such as alternate, decussate, distichous, opposite, radical and whorled, and these are illustrated in Figure 20 and defined in the glossary. Numerous terms describe the shapes of leaves and the incisions in their margins, some of which appear in Figures 21–2.

The angle between a leaf and the supporting stem is termed the **axil**. Most lateral buds arise in leaf axils and these develop into either inflorescences or shoots that become branches. At the base of the leaves of some species is a pair of outgrowths called **stipules** (Fig. 19–20; Pl. 7b) that may be small and insignificant as in *Pultenaea* (bush-pea, Fig. 60) or large and leafy as in roses and *Lathyrus* (sweet pea, Fig. 56). Sometimes the stipules are spines as in *Acacia paradoxa* (hedge wattle).

Leaves are sometimes modified to carry out special functions. For example, the tendrils of sweet pea (Fig. 56) are modified leaflets of the compound leaf, and the succulent leaves of *Mesembryanthemum* have special water-storage tissues. The leaves of *Drosera* (sundew) with their covering of glandular hairs (Pl. 4e) and those of the Venus fly-trap and pitcher plants are special adaptations for trapping insects.

Leaf-like structures that are often smaller than the leaves of the plant are usually referred to as **bracts**. They are commonly found at the bases of branches within an inflorescence or subtending the individual flowers (Fig. 15). Some plants have a very large and conspicuous bract subtending the inflorescence. In the monocots this is often referred to as a **spathe** or spathaceous bract. The large, white, petal-like structure of the arum lily is a spathe, and it subtends the yellow spike-like inflorescence of tiny flowers (Pl. 6b). The brown bud scales that form a protective covering on resting buds, as in *Fraxinus* (ash), and other deciduous and some evergreen plants can also be called bracts. The term **bracteoles** refers to small, bract-like structures that usually occur in pairs, often on the pedicel or calyx of flowers (Figs 61, 97).

Plant Structure and Function

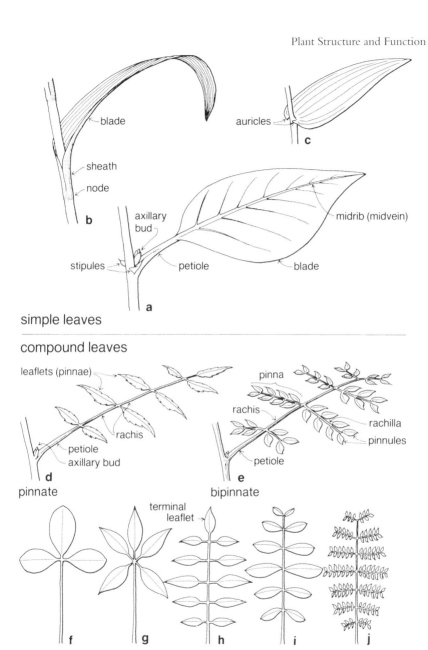

Fig. 19 Leaves—simple and compound: **a** leaf simple, petiolate, stipulate, with reticulate venation; **b** leaf simple, base sheathing, venation parallel; **c** leaf simple, sessile, auriculate, venation parallel; **d–j** compound leaves

45

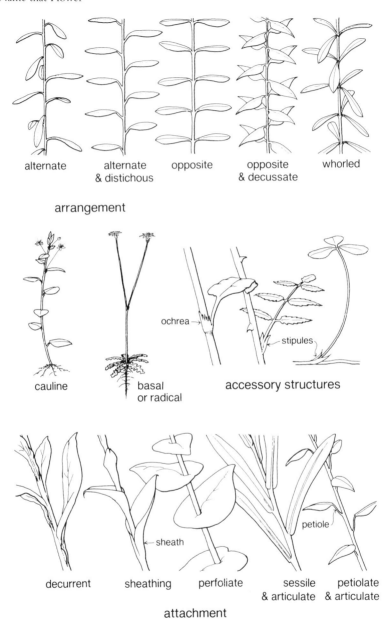

Fig. 20 Leaves—arrangement and attachment, accessory structures

Plant Structure and Function

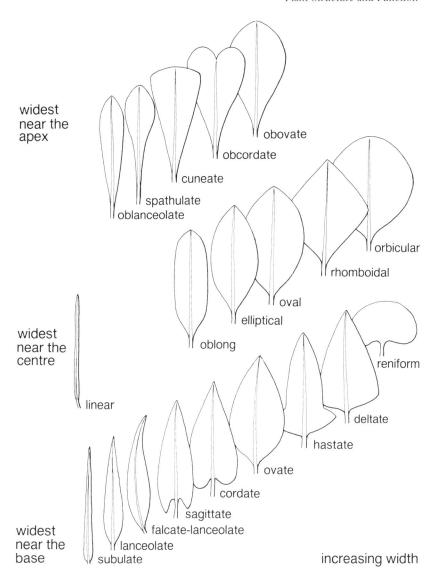

Fig. 21 Leaves—shapes: this diagram covers most of the commonly used terms, but it should be noted that the shapes are not rigidly defined; two or more terms are often used together when considerable variation occurs

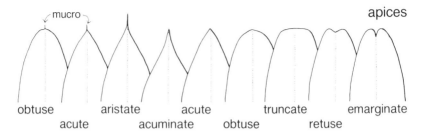

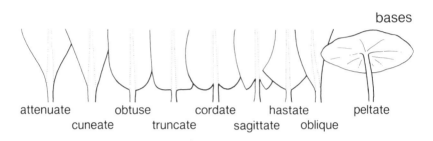

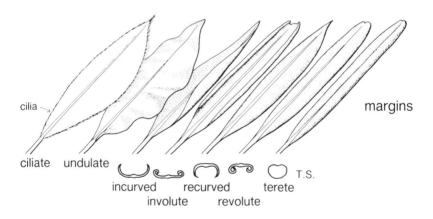

Fig. 22 Leaves—apices, bases, margins

Plant Structure and Function

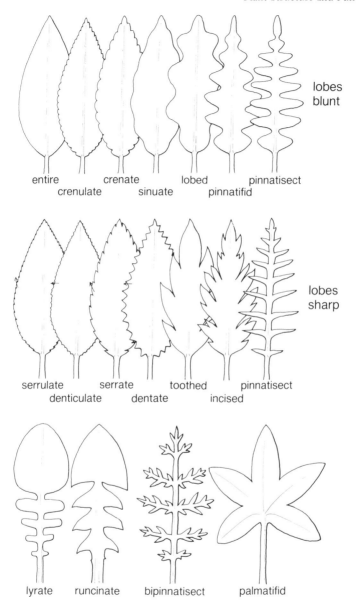

Fig. 23 Leaves—margins: the diagrams are all of simple leaves, although some are deeply dissected

One important function of green leaves is the manufacture of carbohydrates from water and carbon dioxide. The process is called **photosynthesis**, as it is light-dependent, and takes place in the chloroplasts that are present in the cells of special tissues in the leaf. Photosynthesis may be summarised as

carbon dioxide + water + light energy = sugars + oxygen.

The carbon dioxide is taken in from the air through special apertures in the leaf surface called **stomata**, and the water travels up from the roots. The sugars are changed into complex carbohydrates and proteins. The oxygen may be used in plant respiration or pass out into the atmosphere.

Roots

A root system consisting of a tap root and lateral roots is usual in dicots. The importance of the laterals depends to some extent on the life span of the plant, and in trees they are usually large and spreading. Unlike stem branches, lateral roots do not arise from surface buds but are initiated within the root tissue. Many plants, particularly monocots, do not have a tap root but form a mass of relatively small fibrous roots. Such root systems are often added to by the formation of adventitious roots, which originate at the base of the stem (Fig. 105).

The vascular system

Many plant groups including the ferns, pines and their allies, and the flowering plants, possess a specialised system of cells through which water and nutrients move around the plant. The vascular system, as it is called, is composed of two tissues: the **xylem** which is largely water-conducting, and the **phloem** through which the products of photosynthesis are carried. The xylem contains several cell types, including water-conducting cells, and others which provide structural support. All have thickened cell walls and form much of the woody tissue of shrubs and trees. The phloem contains sieve elements, which are thin-walled and act as nutrient-conducting cells, and structural cells may also be present. Both tissues contain storage cells, which are involved in gas and nutrient exchange with the cells around them, and are connected to the sieve elements of the phloem. In woody plants the phloem occurs in a layer just under the bark so that when a tree is 'ring-barked' death is the result of disruption of the food supply to the roots and of the water supply to the aerial parts.

Plant growth

Most of the extension growth of stems and roots takes place near their apices. In zones called meristems, new cells are formed by division, and behind the meristem these differentiate into the various cell types of the stem or root. At the shoot apex, the meristem is more or less dome-shaped and on the flanks of the dome are small bumps called leaf primordia, which ultimately develop into mature leaves. At the root apex the meristem is just behind a mass of cells called the root cap, which protects the meristem as the root pushes down through the soil. A little behind the meristem is the root hair zone where thousands of fine unicellular hairs protrude from the outermost cell layer of the root. Root hairs are relatively short-lived, but are of great importance, as most of the water containing the dissolved minerals required by the plant is absorbed through them.

In the root systems of perennial plants particularly, young roots with their root hairs must be renewed constantly. Some fungi, such as the widespread cinnamon fungus, attack these young roots, and an affected plant usually dies because it can no longer absorb sufficient water.

The increase in girth of woody plants is also due to a meristematic zone. In dicots this is laterally placed, not far under the bark. Some of the cells produced by this meristem add to the xylem, and the others provide new phloem cells, as in most plants the latter do not live for more than one season.

An interesting aspect of plant growth is that new cells and tissues are formed all through the life span of the plant, even in trees that may be a hundred or more years old. The shoot apices on main and lateral branches are the only sources of new leaves so these meristems must be maintained. If the apex of a shoot is destroyed, one or more of the lateral buds further back will usually take over.

6
Classification and Nomenclature —grouping and naming

Within the flowering plants the system of classification is phylogenetic, that is, it is believed to reflect the evolutionary relationships between the groups. However, although it is reasonable to assume that members of the pea family or the orchids have each evolved from a common ancestor, the relationship between their ancestors is a matter of debate. As a result, not all workers agree on the relationships between living plant groups.

Theories about these relationships impact most noticeably on students of plant identification in that they affect the sequence and contents of families assembled in Floras (manuals documenting the plants of a region). Thus in the *Flora of Australia* (bibliography no. 94) and the *Flora of Victoria* (no. 219), the large and diverse families Rosaceae (roses and allies) and Liliaceae (lilies) are both treated in a broadly inclusive sense. By contrast, in the *Flora of New South Wales* (no. 104) plants in these groups are accommodated in a number of smaller families. These differences arise because the Floras are based on classification schemes proposed by different authors.

Quite a number of schemes have been put forward over the last several hundred years. Of the more recent ones, perhaps half a dozen have received most attention in the form of subsequent evaluation and criticism. Despite differences, these major schemes do have a lot in common, for example, when they are translated into linear sequences for inclusion in a Flora, the Magnoliaceae (magnolia family) is usually found early in the sequences. The Asteraceae (daisy family) by contrast is towards the end. This is not to say that daisies have evolved from magnolia-like ancestors, but rather that magnolias and their allies are believed to be more similar to ancestral plants than are the daisies. The members of the daisy family are believed to differ more distinctly (that is to show a greater degree of change) to ancestral flowering plants than do the magnolias.

When discussing the subject of evolutionary change botanists have used the terms 'primitive' and 'advanced' when referring to plant characteristics. A primitive character is one believed to be similar to an ancestral plant feature. Conversely an advanced character is one considered to have evolved such that it differs markedly from ancestral forms. Botanists have been hampered by the meagre fossil record when trying to deduce the features of ancestral flowering plants. Flowers are delicate structures not readily preserved as fossils compared to, say, the woody cones of conifers, or even leaves. In general, characteristics taken to represent a more primitive condition include a woody habit, flowers solitary and symmetrical, with numerous free parts spirally arranged, and a superior ovary. Characteristics often seen as representing an evolutionarily advanced condition include a herbaceous habit, with asymmetrical flowers in more complex (often specialised) inflorescences, with reduced numbers of parts arranged in whorls, and an inferior ovary.

The Asteraceae (daisy family) and the Orchidaceae (orchid family) are two examples of groups with highly specialised flowers and/or inflorescences, compared to say *Magnolia* (Fig. 10) or *Sedum* (Fig. 13). Members of these families are often associated with particular pollinators (frequently insects), and it is believed that the evolution of flowering plants is to a large extent a story of increasing specialisation in the relationships between plants and their pollinators.

The flowering plants are traditionally divided into two major groups, called the dicotyledons (dicots) and the monocotyledons (monocots). These names are based on the fact that the seeds of dicots have two cotyledons or seed-leaves, and those of the monocots have one. The cotyledons are attached to the embryo plant within the seed-coat and are involved in the supply of food to the young plant immediately after the seed germinates. In addition, certain vegetative and floral characters are typical of, but not confined to, each group, as the accompanying chart illustrates.

	Dicotyledons	Monocotyledons
Seed	2 cotyledons	1 cotyledon
Flower	floral parts mostly in 5s, sometimes 4s	floral parts mostly in 3s
Habit	plants woody or herbaceous	almost all species herbaceous
Leaves	venation mostly reticulate	venation mostly parallel
Root system	often a tap root with well-developed lateral root system	usually fibrous

Both dicots and monocots are sub-divided into orders, families, genera and species, the species being the unit of classification. A **species** is recognised as a distinct assemblage of individuals with a unique combination of flower, fruit and vegetative characters and is often thought of as a 'kind of plant'. At one time it was thought that a species could be defined as a group of self-perpetuating individuals that did not breed with other groups, but this is not true of all species as some interbreed or hybridise and give rise to intermediate forms.

Related species are grouped into genera. It is usually clear that the species in a genus are related, as they are often similar in general appearance with a number of features (generic characters) in common. In some cases the characters that separate two similar species are only observed after close examination.

Related genera are grouped into families. However, the characters shared by all members of a family are fewer than the number shared by members of a genus. In spite of this, many families are clearly recognisable.

The Swedish naturalist Linnaeus established the use of the characteristics of the sexual reproductive organs as the basis for grouping plants. He then named them according to the binomial system which gives each species a two-word name. The first word is the name of the genus, or generic name, and the second is the trivial name or specific epithet, which belongs only to one species in the genus. The generic name is written with a capital letter and the specific epithet with a small one. The binomial is underlined in written or typed work and placed in italics in printed publications.

The specific epithet is not restricted to one genus; thus *Eucalyptus alpina*, *Acacia alpina* and *Grevillea alpina* are three distinct species.

After a binomial has been written in full, in subsequent references to the genus it is usual to abbreviate the generic name to the first letter only. For example, a second reference to *Acacia alpina* would be written *A. alpina*, and other species of *Acacia* could be *A. stricta* or *A. suaveolens*.

Sometimes a species is very variable but the variation is insufficient to justify the recognition of additional species. The lower categories that may be used to show this, in order of rank, are subspecies (ssp.), variety (var.), and form (f.). One example is *Banksia ericifolia* var. *macrantha*. The boxed list shows the sequence of the main ranks used in botanical classification.

Kingdom	Plantae (plants)
Division	Anthophyta (flowering plants, Angiosperms)
Class	Dicotyledones (dicots)
Order	Myrtales
Family	Myrtaceae (myrtles)
Subfamily	Leptospermoideae
Tribe	—
Genus	*Eucalyptus*
Species	*E. globulus* (blue gum)
Subspecies	*E. globulus* ssp. *globulus*
Variety	—

A complete classification for *Eucalyptus globulus* (blue gum) showing the main ranks in sequence; many other formal ranks are available (over 20 in all), but not all need to be used for any particular species. Botanists have not reached agreement on the names of the higher ranks and a number of alternatives are used. In the new *Flora of Australia* (bibliography no. 94) the flowering plants are known as 'Magnoliophyta' and the dicotyledons as 'Magnoliopsida'.

An abbreviated word often follows the binomial, as in *Acacia microcarpa* F. Muell. and *Hakea ulicina* R.Br. The addition of these initials to the binomial indicates the person who described and named the species: F. Muell. is short for Ferdinand von Mueller and R.Br. for Robert Brown. The author's name is called 'the authority', and it always appears in 'floras' and official publications. When a name is changed, the authority for the original name is placed in brackets and is followed by the author of the alteration. Therefore, when *Mimosa suaveolens* Sm. was transferred to *Acacia* by Willdenow, the citation became *Acacia suaveolens* (Sm.) Willd. Another authority citation that appears is *Daviesia genistifolia* A. Cunn. ex Benth. The 'ex' connecting the two abbreviations means that Bentham described the species and published the name first proposed, but not published, by Cunningham.

Many names have Latin or Greek roots and often indicate a real or imagined feature of the plant: *Leptospermum* comes from *leptos* = slender, *sperma* = seed; *grandifolium* from *grandis* = large, *folius* = leaf; and *microcarpa* from *micros* = small, *carpus* = fruit.

Other names are associated with famous botanists, explorers or collectors. *Banksia* is named after Sir Joseph Banks, who accompanied Cook on his first voyage. *Brunonia* and the specific epithets *brownianus* and *brownii* all honour Robert Brown, who was the naturalist on Flinders' expedition and who later described and named many plants sent to England by other collectors. Several specific epithets are named after Allan Cunningham, who was sent to Australia by Banks to collect plants for the Kew Botanic Gardens, and accompanied Oxley and King on their expeditions. Cunningham made many other exploring trips in Australia, New Zealand and Norfolk Island, and later became Superintendent of the Sydney Botanic Gardens. He described and named much of his material, and the authority 'A. Cunn.' appears regularly in our floras. For dictionaries of plant names and their meanings see bibliography numbers 15 and 204, and for two accounts of the people involved in early Australian botany, see bibliography numbers 43 and 44.

Cultivated or garden plants also have botanical names. Some are popularly known by their generic name, others by common name. Examples include *Zinnia elegans* Jacq., *Antirrhinum majus* L. (snapdragon), *Papaver nudicaule* L. (Iceland poppy) and *Quercus robur* L. (English oak). The authority 'L.' in these examples stands for Linnaeus.

With a few exceptions, family names are derived from a genus in the family, which then becomes the 'type' genus. Proteaceae comes from the South African genus *Protea*, and the genera *Banksia, Hakea, Grevillea, Lomatia* and *Isopogon* are some Australian members of the family. Some families were given names associated with obvious characteristics, so Compositae described the inflorescence of the daisy group and Labiatae the two-lipped flowers prevalent in the mint family. It has been recommended that Compositae should be changed to Asteraceae, from the genus *Aster*, and Labiatae to Lamiaceae from the genus *Lamium*, and that all family names should end in 'aceae' (usually pronounced '-*ay-see*'). Related families are grouped into orders, the names of which end in 'ales' (usually pronounced '-*ay-lees*').

Name changes are a common cause of irritation to anyone working with plants. They indicate the continuing efforts to refine the classification system. One reason for a change is the discovery, during the revision of a family or genus, of an earlier name and description of a plant than the one currently in use. The *International Code of Botanical Nomenclature* (bibliography no. 99) establishes the rules for naming plants, one of which gives priority to the first-published name. Sometimes plants were classified incorrectly, often as a result of the work being done with insufficient material, and subsequent revision requires a change of name. The delimitation of families is often a matter of opinion or interpretation, so a genus that does not quite fit in one family may be moved into a closely related one. All changes must be notified in a scientific journal and, if generally accepted, will be incorporated into floras.

Cultivated plants originating from a selection process (which may also involve hybridisation) are known as cultivars (cv.) and the names of these are distinguished by a cultivar epithet, which is enclosed in single quotation marks. Cultivars from *Prunus persica* (peach) include *P. persica* 'Magnifica' (double-white peach) and *P. persica* 'Roseoplena' (double-pink peach). Some cultivars are the result of selection from wild populations and their geographic origin may be suggested by the cultivar epithet, for example, *Grevillea confertifolia* 'Major Mitchell' is a form from the plateau of that name in the Grampians, Victoria. Preferred selections may be perpetuated by taking cuttings, as is done with grevilleas, or grafted on to another stock plant, which is the method for most fruit trees.

Hybrids are shown with a multiplication sign either between the names of the two parents, if they are known, or before the specific epithet if the hybrid has been formally named. *Magnolia* × *soulangeana* is a hybrid between *M. denudata* and *M. liliflora*.

In addition to their botanical name the majority of plants have a common name in the language of the country or state in which the plant grows. However, as common names are not universal and may change from region to region, their use often leads to errors and confusion. Use of the binomial is recommended as it eliminates many mistakes in identity and is international in application.

Detailed descriptions of families, genera and species are assembled in 'Floras'. These are commonly compiled on a regional basis and Australia has a number of state Floras as well as some for specific regions such as Central Australia, in the broad sense, and the Sydney region. *A Flora of Australia* by Bentham was published in 1860 but a new Flora is now in preparation with a number of volumes already published (see bibliography no. 94). There are also specialist books on families such as the orchids, and on the larger genera such as *Eucalyptus* and *Banksia*.

For further reading on the classification of flowering plants see bibliography numbers 94, 136, 143, 181 and 182, and for plant nomenclature see numbers 96, 99, 103, 150 and 215. *Plant Names* (no. 150) provides a useful introduction to the nomenclature of cultivated plants.

7

The Process of Identification

The steps involved in formally identifying a plant, when you have no prior knowledge of its identity, involve
- observing the plant's characteristics, particularly its floral structure
- working through an appropriate botanical key—you will need to know whether the plant is growing wild or is in cultivation because most reference books are restricted in content along these lines.

Equipment

Minimum requirements: a single-sided razor blade; two darning needles, pushed into corks for handles; a pair of tweezers; a 10× hand lens, or a lens or magnifying glass mounted on a stand. Refinements: one or two pairs of jeweller's forceps with fine points; a binocular dissecting microscope.

Fig. 24 Using a hand lens: hold the lens close to the eye and bring the plant up until it is in focus

The Process of Identification

Choosing the flower to look at

As a means of gaining experience in the process of identification a good place to start may be to examine a flower that is illustrated in this book and then compare your observations with the drawings and captions. This means that you will already know the plant's name which can help to guide your path through the key. It is also a good idea to start with a large flower that has a simple structure, to get the feel of the process of dissection. Beware of horticultural forms with multiple petals, such as roses, carnations and some fuchsias, and avoid daisies at first.

Some plant species are very variable, so it is unwise to assume that all flowers on a plant, or in a population, will be identical. Minor variations may not be important, but make sure you select representative flowers. Stamens commonly open in sequence, and the early ones may fall before the last one opens, so if you are in doubt about the number of stamens, look at a bud. On the other hand, older flowers often show the ovary structure more clearly. Collect fruits, if available, and note habit and leaf characters.

Interpreting what you see

Remove the flower parts carefully and lay them out. Note the following points:

Perianth	number of whorls (one or two), or parts arranged spirally
	presence of sepals as well as petals
Calyx	number of sepals (they may be very small)
	free or united
	other features (hairs, dissected margins . . .)
Corolla	number of petals
	free or united
	other features (colour, hairiness—internal or external . . .)
Androecium	number of stamens (1–10 or numerous)
	free or united
	epipetalous or joined to the receptacle
	other features (appendages, hairiness . . .)
	presence of staminodes
Gynoecium	carpels free—number of carpels (1–10 or numerous)
	carpels united—
	ovary superior or inferior
	number of carpels—decide by noting:
	stigma simple or lobed
	style simple or branched
	number of loculi in the ovary (cut a T.S.)
	if one loculus, number of placentas
	placentation (cut a T.S. and L.S., see also Fig. 25; Pl. 2f)

Name that Flower

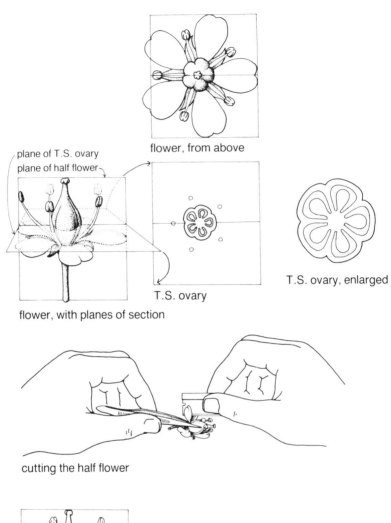

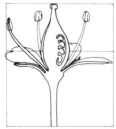

Fig. 25 Flower sections: cutting a flower to examine its structure

The floral formula for this hypothetical example is K5 C5 A5 G(5). Similar sections have been used in Figs 10–14, and 26–131. See also Pl. 2f.

Cutting a section

TRANSVERSE SECTION (T.S.)

Identify the ovary, hold the flower by the stalk and make horizontal cuts. It is easier to see the loculi if you cut a thin slice, and examine that with a lens as it sits on the edge of the razor blade. If the flower is small and the ovary superior, it may be better to hold the flower by the petals and start cutting just below the calyx.

LONGITUDINAL SECTION (L.S.)

Examine the outside of the ovary for signs of its structure. Often there will be lines or creases indicating the positions of the loculi. Try to cut through the centre of at least one loculus. If the cut is a little off centre it may be difficult to interpret what you see.

When cutting a half flower, and the flower is actinomorphic, it may be cut in any plane, but often a carefully judged half flower will also give a good section of the ovary. If the flower is zygomorphic, choose the plane that will result in more or less equal halves. Lay the flower on its side and cut carefully but firmly (Fig. 25).

Using keys

You will now need a book (this will usually be a Flora), appropriate for your region, that includes a key to the families of plants. The bibliography lists those available for the Australian flora, and those by Bailey (no. 13), Spencer (no. 209), and Walters (no. 220) that are useful for garden plants.

Keys provide a method of identifying a plant by the process of elimination. They are usually quite artificial in the sense that most keys do not take account of real relationships, but only of characters that will separate one species from another. The first key steps in Floras usually separate non-flowering plants such as ferns and conifers from the flowering plants. If your sample is a flowering plant, the next steps will often separate monocots and dicots, and then divide the families into groups using contrasting character pairs, such as 'perianth of one or two whorls', 'petals free or united', or 'ovary superior or inferior'. These pairs are termed couplets, and each of the character statements in a couplet is a lead. In practice, there may be three or four leads in some couplets, a situation that is often unavoidable and does not necessarily complicate the issue. The two common types of keys, bracketed and indented, differ only in layout. In bracketed keys the leads of a couplet are set out together. Using the character pairs mentioned above, a possible bracketed key may be:

1	Perianth of two whorls of parts ...(go to couplet 2)	2
	Perianth of one whorl of parts ....(go to couplet 3)	3
2	Petals free..……………..............	4
	Petals united ....…………….	5
3	Perianth parts free ....……………….............	6
	Perianth parts united..…………….....	7
4	Ovary superior ....…………........ Group A
	Ovary inferior ....……….. Group B
5	Ovary superior..……............. Group C
	Ovary inferior...…...…..................... Group D
6	Ovary superior ....…….....….............. Group E
	Ovary inferior..…................... Group F
7	Ovary superior ....………..…..….............. Group G
	Ovary inferior...… Group H

An indented key using the same pairs of characters would be set out as follows:
    1 Perianth of two whorls of parts
        2 Petals free
            3 Ovary superior... Group A
            3 Ovary inferior..…... Group B
        2 Petals united
            4 Ovary superior ....…………. Group C
            4 Ovary inferior ....…….. Group D
    1 Perianth of one whorl of parts
        5 Perianth parts free
            6 Ovary superior ....…….. Group E
            6 Ovary inferior..…….. Group F
        5 Perianth parts united
            7 Ovary superior ....…….. Group G
            7 Ovary inferior..….. Group H

The Process of Identification

Both methods split the families into eight groups with the following combinations of characters:

A Perianth of two whorls, petals free, ovary superior
B　　　　"　　　　　　free　"　inferior
C　　　　"　　　　　　united　"　superior
D　　　　"　　　　　　united　"　inferior
E Perianth of one whorl, parts free, ovary superior
F　　　　"　　　　　　free　"　inferior
G　　　　"　　　　　　united　"　superior
H　　　　"　　　　　　united　"　inferior

Thus, in the first couplet of the bracketed key, a flower in group A would take the first lead, and progress to couplet two, indicated by the number in the right-hand column. In this couplet, the specimen again matches the first lead and progresses to couplet four. At couplet four the correct choice, ovary superior, leads to group A.

At the beginning of an indented key to a large number of families or genera, the first and second leads of a couplet may not be on the same page, so check this before proceeding. In the example above, the two opposing leads in each couplet are marked with the same number; sometimes letters or other symbols are used, for example A and AA, A and A', A and *A etc. A plant in group G would take the second lead of couplet one. From here the next step is to couplet five. The second lead again applies and in couplet seven the first choice leads to group G.

Some sections of representative keys are reproduced on pp. 64–7 to illustrate the way in which authors have tackled the setting out. These keys are included here only as examples, and the books in which they have been published may be out of print or superseded by more recent editions. For those who are interested to follow the keying-out process through other published keys, routes through keys are set out for some of the plants illustrated in this book. (See the introduction to dicot families, Chapter 8.)

Key 1 is taken from the *Handbook to Plants in Victoria*, by J. H. Willis (pp. 326, 332–3), and includes the key to genera in the family Rutaceae, and the key to species of *Eriostemon*. *Eriostemon myoporoides* (which is now called *Philotheca myoporoides*, Fig. 65) takes the following route:

(p. 326) 1-2-3-6-7-8-9- *Eriostemon* (p.332) 1-2-3-4- *E. myoporoides*

Willis' key to species is a bracketed one and follows the usual pattern except that the names of species, and extra information, have been inserted at relevant places. Thus the first lead in couplet four relates to *E. myoporoides*, which follows immediately. The alternative lead starts with the same word and is prefixed with a dash.

Name that Flower

Family RUTACEÆ

alternative leads of couplet 1

1. Fruit *indehiscent, succulent*, whitish (glabrous tree of E. Gippsland, with flat, oblong, opposite *or* alternate leaves 2-4" long and small slenderly stalked flowers in axillary cymes) *Acronychia* (p. 341)
   Fruit *dehiscent, dry*, lobed 2
2. Flowers *minute, sessile, in terminal heads*; carpels 2; leaves alternate, terete, <8 mm. long (ericoid shrubs of Mallee)
   Microcybe (p. 337)
   Flowers *conspicuous, not capitate*; carpels normally 4 or 5 3
3. Leaves *alternate, simple*; petals 5 6
   Leaves *opposite*, sometimes compound; petals 4 4
4. Petals *valvate, cohering* in a cylindrical tube, 1-5 cm. long; calyx mostly cup-shaped; leaves simple, broad ($\pm$ stellately woolly, rarely glabrous shrubs) *Correa* (p. 338)
   Petals *free* as soon as buds open, <1 cm. long 5
5. Stamens 8; disk of corolla *entire* *Boronia* (p. 327)
   Stamens 4; disk *with 4 gland-like lobes* *Zieria* (p. 330)
6. Stamens 5 (rare N.W. tree of Murray Valley, with linear leaves 3-6" long and very small whitish flowers) *Geijera* (p. 327)
   Stamens 10 (habit shrubby) 7
7. Calyx *minute*, hidden amongst hairs and quite *inconspicuous*; petals *yellow*; disk *absent*; stigma *large* *Asterolasia* (p. 337)
   Calyx *conspicuous* or, if very small, then *not* hidden by hairs; disk *present*; stigma *small* 8
8. Stamens *glabrous, spreading* at anthesis; petals *valvate* in bud, mostly *yellowish*, often scaly, 4-6 mm. long *Phebalium* (p. 333)
   Stamens $\pm$ *ciliate* or even woolly, *erect or incurved* at anthesis; petals usually imbricate in bud, *white or pink*, never scaly, 6-12 mm. long 9
9. Anthers with only a *minute* appendage; leaves and branches often manifestly *tuberculate-glandular* *Eriostemon* (p. 332)
   Anthers tipped by a *prominent bearded* appendage; leaves and branches *never* prominently glandular *Crowea* (p. 333)

Key 1 Bracketed key to the Victorian genera of the family Rutaceae, from *A Handbook to Plants in Victoria* by J. H. Willis (Melbourne Univ. Press, 1972, by permission)

Key 1 (cont'd) Bracketed key to the Victorian species of *Eriostemon*, from *A Handbook to Plants in Victoria* by J. H. Willis (Melbourne Univ. Press, 1972, by permission)

332   RUTACEÆ

ERIOSTEMON Sm. (1798)

alternative leads of couplet 1

1. Leaves *pungently pointed*, smooth, 10-15 mm. long; flowers *solitary* in axils, white (rosy-red in bud), 4-6 mm. long, gland-dotted (low heath-like bush of far west and southern Mallee):

E. pungens Lindl. in Mitch. *Three Exped. E. Aust.* 2: 156 (1838).
Phebalium pungens (Lindl.) Benth. *Flor. aust. 1*: 338 (1863).
Illust.: Cochrane, Fuhrer, Rotherham & Willis, *Flowers & Plants Vict.* t. 132, col. (1968).
Vern.: Prickly Wax-flower. *Distr.:* BCDFGJ—also S.A., N.S.W.

alternative leads of couplet 2

— —Leaves *never* pungent; flowers often in clusters   2
2. Leaves ± 1 mm. wide   5
— — — Leaves >2 mm. wide   3

3. Leaves *obovate*, 5-10 mm. long, retuse *or* the apex recurved, ± glaucous, thick, with inconspicuous midrib; flowers showy, pedicellate, solitary or sometimes paired on a short peduncle (straggling, chiefly western shrub):

authority

E. verrucosus A. Rich. in *Voy. l'Astrolabe (Bot.)* 2: 74, t. 26 (1834).
E. obovalis sens. Ewart *Flor. Vict.* 705 (1931) atque auctt. plur., *non* A. Cunn. (1825).

Illust.: Richard (*l.c.*); Cochrane, Fuhrer, Rotherham & Willis, *Flowers & Plants Vict.* t. 324, col. (1968); Galbraith, *Wildflowers Vict.* ed. 3: t. 83 (1967); Brooks, *Aust. native Plants* t. inter 64 & 65 (1959), as *E. verrucosa*.
Vern.: Fairy Wax-flower. *Distr.:* CDHJMNSW—also S.A., Tas., N.S.W.

—Leaves *oblong to oblanceolate*, mostly >2 cm. long, flat, the apex ± mucronate (tall shrubs of E. highlands)   4

lead of couplet 4 relevant to yoporoides

4. Flowers *several*, in an axillary umbel with conspicuous peduncle; carpels ± *beaked*; leaves broadly *oblanceolate* to linear-lanceolate, pointed, 1-4" long:

E. myoporoides DC. *Prodr. 1*: 720 (1824).

list of references to Illustrations

Illust.: Cochrane, Fuhrer, Rotherham & Willis, *Flowers & Plants Vict.* t. 429, col. (1968); Ewart, *Flor. Vict.* fig. 283 (1931); Galbraith, *Wildflowers Vict.* ed. 3: t. 84 (1967); Burbidge, *Flor. Aust. Cap. Terr.* fig. 233 (1970).
Vern.: Long-leaf Wax-flower. *Distr.:* NPRSVWZ—also N.S.W., A.C.T., Qd.

distribution

—Flowers mostly *solitary* but, if ever 2-3 in axils, then *without* a common peduncle; carpels *never* beaked; leaves oblong to oblanceolate, obtuse, mostly <1¼" long (tall shrub or tree of E. Gippsland):

E. trachyphyllus F. Muell. in *Trans. phil. Soc. Vict. 1*: 99 (1855).

Illust.: Pescott, *Wild Life (Melb.) 1*: 24 (Oct. 1938).

vernacular (common) name

Vern.: Rock Wax-flower. *Distr.:* SVWZ—also N.S.W.

5. Leaves 1-4 mm. long, *straight*, ± claviform, *obtuse*, with a few *large tubercles*; flowers *terminal*, solitary or few together, 4-6 mm. long:

65

> RUTACEÆ 333
>
> **E. difformis** A. Cunn. ex Endl. et al. *Enum. Plant. Hueg.* 15 (1837).
>   *E. gracilis* sens. Ewart *Flor. Vict.* 705 (1931), *non* R. Graham (1834).
> *Illust.:* Galbraith, *Aust. Plants 1*[5]: 7 (1960).
> *Vern.:* Small-leaf Wax-flower. *Distr.:* ABCDHJM—also W.A., N.S.W., Qd.
>
>   [Corollas are typically hairy on the outside, but Grampians and some Mallee populations have *glabrous* petals (except for occasional marginal cilia).
>   The *E. gracilis* described by Graham in *Edinb. new phil. J. 16*: 175 (1834) was almost certainly referable to N.S.W. *Philotheca salsolifolia* (Sm.) Druce—*teste* P. G. Wilson.
>   In *Nuytsia 1*[1]: 31 (1970), P. G. Wilson has segregated and described as a new species *E. angustifolius*, differing from *E. difformis* in the *glabrous* outer surfaces of its petals, slightly longer anthers (1-1.5 mm.) but shorter cocci (± 3 mm.). This taxon is often co-extensive with *E. difformis* over the southern part of the latter's range (CDM), including Bendigo district.]
>
>   —Leaves 10-15 mm. long, $\pm$ *upward-curving, acutely pointed*, thick, *concave beneath*; flowers *axillary* and solitary (extremely rare plant of Myrniong district near Bacchus Marsh, where perhaps now extinct):
>
> **E. scaber** Paxton in *Paxton's Mag. Bot. 13*: 127 (1846).
> *Illust.: Paxton's Mag. Bot. 13*: t. opp. 127, *col.* (1846).
> *Vern.:* Rough Wax-flower. *Distr.:* N—also N.S.W., Qd.

Key 1 (cont'd) Bracketed key to the Victorian species of *Eriostemon*, from *A Handbook to Plants in Victoria* by J. H. Willis (Melbourne Univ. Press, 1972, by permission)

    Key 2 is an indented one to genera in the family Epacridaceae, from the *Flora of the Sydney Region*, by Beadle *et al.* Letters are used instead of numbers to indicate the leads in each couplet. *Leucopogon* (Fig. 90 or 91) takes the following route:

               *A-C-*D-E- *Leucopogon*

The numbers on the right-hand side indicate the sequence of genera in the pages following the key.

Points to note

It is important to read both or all leads in the couplet before deciding which one to follow. Not all couplets are as clear-cut as our examples at the beginning of the previous section on using keys and, further on in a key, couplets tend to become more complex.

    It is useful to write down the steps you have taken as you go through a key. It is then easier to check your route if an error is made. If you do not arrive at a satisfactory answer, it is likely that the mistake is yours, but it is important to note that keys are not infallible. Sometimes a species varies so much that it is

The Process of Identification

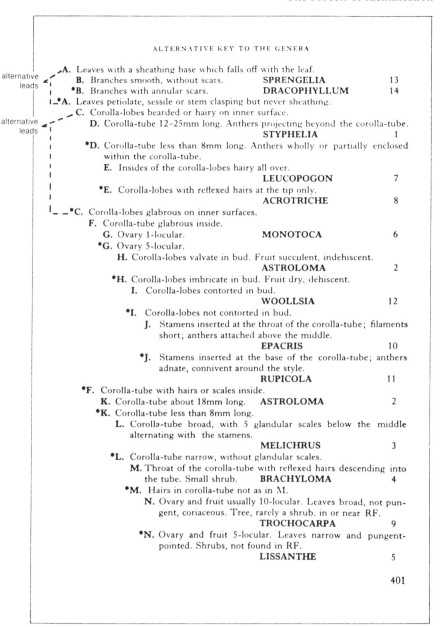

ALTERNATIVE KEY TO THE GENERA

alternative leads
- **A.** Leaves with a sheathing base which falls off with the leaf.
  - **B.** Branches smooth, without scars. **SPRENGELIA** 13
  - ***B.** Branches with annular scars. **DRACOPHYLLUM** 14
- ***A.** Leaves petiolate, sessile or stem clasping but never sheathing.

alternative leads
  - **C.** Corolla-lobes bearded or hairy on inner surface.
    - **D.** Corolla-tube 12–25mm long. Anthers projecting beyond the corolla-tube. **STYPHELIA** 1
    - ***D.** Corolla-tube less than 8mm long. Anthers wholly or partially enclosed within the corolla-tube.
      - **E.** Insides of the corolla-lobes hairy all over. **LEUCOPOGON** 7
      - ***E.** Corolla-lobes with reflexed hairs at the tip only. **ACROTRICHE** 8
  - ***C.** Corolla-lobes glabrous on inner surfaces.
    - **F.** Corolla-tube glabrous inside.
      - **G.** Ovary 1-locular. **MONOTOCA** 6
      - ***G.** Ovary 5-locular.
        - **H.** Corolla-lobes valvate in bud. Fruit succulent, indehiscent. **ASTROLOMA** 2
        - ***H.** Corolla-lobes imbricate in bud. Fruit dry, dehiscent.
          - **I.** Corolla-lobes contorted in bud. **WOOLLSIA** 12
          - ***I.** Corolla-lobes not contorted in bud.
            - **J.** Stamens inserted at the throat of the corolla-tube; filaments short; anthers attached above the middle. **EPACRIS** 10
            - ***J.** Stamens inserted at the base of the corolla-tube; anthers adnate, connivent around the style. **RUPICOLA** 11
    - ***F.** Corolla-tube with hairs or scales inside.
      - **K.** Corolla-tube about 18mm long. **ASTROLOMA** 2
      - ***K.** Corolla-tube less than 8mm long.
        - **L.** Corolla-tube broad, with 5 glandular scales below the middle alternating with the stamens. **MELICHRUS** 3
        - ***L.** Corolla-tube narrow, without glandular scales.
          - **M.** Throat of the corolla-tube with reflexed hairs descending into the tube. Small shrub. **BRACHYLOMA** 4
          - ***M.** Hairs in corolla-tube not as in M.
            - **N.** Ovary and fruit usually 10-locular. Leaves broad, not pungent, coriaceous. Tree, rarely a shrub, in or near RF. **TROCHOCARPA** 9
            - ***N.** Ovary and fruit 5-locular. Leaves narrow and pungent-pointed. Shrubs, not found in RF. **LISSANTHE** 5

401

Key 2 Indented key to the genera of the family Epacridaceae in the Sydney region, from *Flora of the Sydney Region* by Beadle et al. (Reed, Sydney, 1972, by permission)

67

almost impossible to cope with the whole range when writing a key. Most keys are written for the plants of a particular region, and some weeds and most garden plants may not be included. Occasionally such a plant may key out easily to an incorrect answer, so it is important to check identifications against written descriptions or illustrations. Numerous plant images are now on the internet, which can be a very useful source of information. When accuracy is important, however, it is perhaps unwise to accept the identity of internet images without checking against an authoritative text. A small selection of websites is given at the end of the bibliography.

The members of some families show considerable variation of form and it may not be possible to 'key them out' as a single group. Thus, a family may be brought out several times, as some groups of genera follow different routes. Similarly, in a key to genera, a particular genus may appear more than once. For an example, see the family key in Willis (bibliography no. 233), where Rutaceae keys out several times; some of the routes are shown at the end of the section on this family in Chapter 8.

Computer keys

Electronic keys to identify plants (for use on a computer) are similar in principle to traditional printed keys but have one significant advantage—the user is not required to start from the beginning and then work through the key following a particular order of couplets. With computer keys, information concerning the characteristics of the sample to be identified can be entered into the computer in any order. As this information is entered, the computer systematically searches its memory of the features of the various plant groups, and rejects those that do not match. This leaves a progressively smaller number of possible 'matches'.

Ideally there will be only one final answer, but if the information entered is limited (for example if your sample does not have flowers and you are relying on vegetative features only) then the computer will provide a list of possible answers. This is an advantage over traditional printed keys which are limited by the quality of the specimen to be identified. For example, if your sample has flowers but no fruit, and the next couplet in the key requires information about the fruit, then progress is either halted, or you must follow both leads in the hope that the correct path will become clear later in the key.

Remember that the user must still correctly interpret the characteristics of the sample to be identified. To assist with this process computer keys often provide guidance, usually in the form of small diagrams, explaining the use of descriptive terms.

The Process of Identification

Additional features are often included with computer keys such as the ability to select the 'best' characteristics to enter that will most effectively divide remaining groups, and the option to view the similarities and differences between selected groups. Several electronic keys are listed towards the end of the bibliography.

8

Plant Families

Dicotyledons

One of the criteria used in selecting the families to illustrate this book was that they should include most of the common genera in the vegetation of south-eastern Australia. The smaller families, represented by only one or two genera or species, were included either because their bright flowers make them visually abundant, for example *Hibbertia* (Fig. 75; Pl. 1b), or because the interpretation of their floral structure is difficult, as illustrated by *Sarcocornia* (Fig. 37). Others have interesting features such as insect-trapping leaves or a parasitic habit.

The arrangement of families closely follows the sequence used by Willis in *A Handbook to Plants in Victoria*, which recognises relationships between plant groups. While this is not the only such sequence proposed, it has been widely used. As far as is possible in a linear sequence, related families are placed together.

For each family, brief notes are provided by way of introductory background. For the larger families, the characteristic floral structure is set out in a way that allows ready comparison with other families. This covers the usual variation and may not cope with the possible extremes included in a family. Consulting the illustrations will help you to visualise the structure and bring it to life. For the smaller families, the discussion concentrates on the illustrated examples.

The 'spotting characters' are features, usually readily observed, that are characteristic of a particular group. They may not all be present in all members of a group, but are sufficiently common to be a useful guide, particularly in the field.

The illustrations and captions should provide enough detail to allow the examples to be used for practice with the keys published in the various regional Floras (see bibliography). The floral formulae used in the captions to convey basic floral details are explained at the end of Chapter 2.

Dicotyledons

For those who have access to one or more of the texts by Beadle *et al.* (bibliography no. 18), Black (no. 24), George/Orchard (no. 94), Jessop (no. 127), Willis (no. 233) or Carolin and Tindale (no. 42), we have set out the way in which some of the plants we have illustrated could be traced through the keys provided in these books. The correct routes are given using the letters or numbers of the appropriate key, but these will only make sense if you have the book in front of you. This information will be found at the end of the sections on the families Proteaceae, Rutaceae and Epacridaceae and for the genera *Sarcocornia* (Fig. 37), *Acaena* (Fig. 45) and *Thomasia* (Fig. 74). It is hoped that this will encourage greater use of these standard texts. From our experience, many people own copies of keys but tend to avoid using them.

The following list attempts to rank some of the families dealt with in terms of the simplicity of their floral structure. It is intended as a guide to where to start (or where not to start) in approaching plant families.

Structure straightforward	*Structure unusual or complex*
Dilleniaceae	Asteraceae
Liliaceae	Brunoniaceae
Pittosporaceae	Casuarinaceae
Tremandraceae	Chenopodiaceae
	Euphorbiaceae
	Orchidaceae
	Poaceae
	Proteaceae
	Rhamnaceae
	Stylidiaceae
	Thymelaeaceae

For further reading about plant families, see numbers 13, 32, 57, 63, 109, 110, 140, 143, 164, 152, 221 and CD6 in the bibliography.

CASUARINACEAE she-oaks

This family of shrubs and trees is widely distributed in Australia and extends to South-east Asia and the Pacific Islands. The trees commonly have a drooping habit with green, slender, grooved and jointed branchlets. At the joints are whorls of small, dark-coloured, scale-like leaves; the number of scale-leaves per whorl is used in the delimitation of species.

The she-oaks are either monoecious or dioecious. The male flower has 2 scale-like perianth parts and 1 stamen. Each flower is subtended by a bract and the flowers are borne in elongated spikes. The female flower has no perianth parts and the 2 fused carpels, with 2 long reddish styles, are subtended by a bract and 2 tiny bracteoles. The flowers form small globular heads and at maturity the bracts become woody to form the characteristic woody cone. The flowers are wind-pollinated.

There are two genera in southern Australia, *Casuarina* and *Allocasuarina*. The members of the latter genus were formerly included in *Casuarina* (Pl. 4c–d). For descriptions and keys to Australian species see *Flora of Australia* vol. 3 (bibliography no. 94).

PROTEACEAE protea family, banksias, grevilleas, hakeas

Most of the 79 genera in the Proteaceae are found in the southern hemisphere: 46 occur in Australia and the remainder are mainly South African or South American. In spite of superficial morphological similarities, the Australian and South African genera are not closely related and none is common to both regions. The family is a very old one and the distribution patterns are thought to have existed before the separation of the southern land masses. There is great diversity of form within the Proteaceae, but its floral structure is distinctive. The family name is derived from the South African genus *Protea*, often grown in Australia as an ornamental.

Many species of *Banksia*, *Hakea* and *Grevillea* produce abundant nectar, which was utilised by the Aboriginal people, who either sucked the flowers or soaked them in water to produce a sweet drink. Timber from two rainforest trees, *Grevillea robusta* (silky oak, Figs 31, 32) and *Cardwellia sublimis*, is used for furniture. Two species of *Macadamia* produce the macadamia nuts of commerce, and *Telopea* (waratah), *Banksia* and *Protea* are cultivated for the cut-flower trade. Many members of the family are planted as ornamentals, including the genera *Hakea*, *Grevillea*, *Banksia*, *Telopea*, and *Stenocarpus* (firewheel tree) from Australia and *Protea*, *Leucodendron* and *Leucospermum* from South Africa. Grevilleas hybridise readily and the many cultivars available are planted extensively in parks and gardens.

PROTEACEAE

Members of the family are illustrated in Figures 26–36 and Plates 2a–b. For further reading on various genera see the list at the start of the bibliography. For descriptions and keys to Australian species see *Flora of Australia* vols 16, 17A, 17B (bibliography no. 94).

FLORAL STRUCTURE

Flowers
: Either actinomorphic or zygomorphic. Usually bisexual, occasionally unisexual. Inflorescences complex, often racemose, cone-like, spike-like or in dense heads. Flowers occasionally solitary and axillary (*Persoonia*).

Perianth
: Tepals 4, petaloid, free or united. In bud the tepals are valvate and more or less united, at anthesis they either roll back and become free, or remain united at the base.

Androecium
: Stamens 4, sometimes 3, usually epitepalous. Filaments often very short, so the anthers appear to be joined directly to the tepals (Figs 33, 35; Pl. 2a). In some genera, as in *Conospermum* (smoke-bush), 1 anther cell or 1 or more anthers may be sterile (Fig. 30).

Gynoecium
: Carpel 1, ovary superior, stalked or sessile. Ovules 1–many. Style-end often conical or discoid. Hypogynous glands usually present, but variable in shape and number, reduced to one in *Grevillea* and *Hakea* (Figs 32, 33, 35; Pl. 2a). In many genera, before the flower opens, the pollen is shed on to the style-end, sometimes called the pollen-presenter (Fig. 33; Pl. 2b). The stigmatic area, in the middle of the pollen-presenter, becomes receptive some time after anthesis, and when the pollen from that flower has dispersed.

Fruit
: Either a woody or leathery follicle, or a drupe (*Persoonia*) or a small nut (*Conospermum, Isopogon*).

Members of the family are shrubs or trees. Leaves are mostly simple but often lobed or deeply divided, and usually stiff and leathery. Many genera have tough, terete, pungent leaves. There are no stipules. Young leaves and branches are often bronze in colour due to the presence of T-shaped hairs, which are usually shed as the part matures.

SPOTTING CHARACTERS

Perianth single, 4-partite, often characteristically zygomorphic, with 4 epitepalous stamens. Habit woody, leaves leathery, often terete and pungent. Fruit often a woody or leathery follicle.

Plant Families

ROUTES THROUGH KEYS

Routes through several keys are given here for members of the family Proteaceae illustrated in this book. Texts are referred to by author's name and bibliography number. Note that what is now *Hakea decurrens* was previously included within *H. sericea*, and will key to that species in older books.

BEADLE *et al.* (18)
Key to families, p. 99.
A-Dicots
 -*A-*B-*C-*D-E-F-H-I- group 2 -*A-*B-E- Proteaceae
 -*A-*B-*C-*D-*E-L-*M-N-O- group 10 -*A-*C-*D-G-*H- Proteaceae

Key to genera, p. 212.
A-B-*C-F- *Banksia* (p. 218)-*A-E-F- *B. marginata*
A-*B-G-H- *Persoonia*
A-*B-G-*H- *Hakea* (p. 220)-A-B-*C-*E-F- *H. sericea*
A-*B-*G-I-*J-K-L- *Grevillea*

BLACK (24)
The main key to families (p. 23) can hardly be used for identification. The families are keyed out more or less in the sequence in which they appear in the text, and each appears only once in the key. Thus, much variation must be compressed into a small space. Some couplets have numerous leads, which introduces further complexity. The key is useful to show the type of characters used in the classification of the higher ranks, (families, orders, etc.) and emphasises the difficulties in providing clear-cut distinctions between them.

Key to families, p. 23. Alternative leads in a couplet are here designated A1, A2, A3; B1, B2, etc.
A2-B2-C2- subclass 1 (p. 24) -A2-B1- Proteaceae

Key to genera, p. 262.
A1-B1-C2- *Adenanthos*
A1-B2- *Conospermum* - *C. mitchellii*
A1-B2- *Persoonia* - *P. juniperina*
A2-D1- *Hakea*
A2-D1- *Banksia* - *B. marginata*
A2-D2- *Grevillea*

PROTEACEAE

JESSOP AND TOELKEN (127)
Key to families, vol. 1, p. 65. The second lead of each pair is here designated by ' following the number.
1'-3'-4-Dicotyledonae (to p. 66)
1'-2'-3'-8'-9-10'-32'-33'-43'-47'-53-54-55'-56'-58- Proteaceae

Key to genera, p. 120.
1-2'-3- *Banksia* (p. 122) 1- *B. marginata*
1'-4- *Conospermum*
1'-4'-5- *Persoonia* (p. 153) - *P. juniperina*
1'-4'-5'-6- *Adenanthos* (p. 121) 1'- *A. terminalis*
1'-4'-5'-6'-7- *Grevillea* (p. 125) 1'-7'-14'-15'-21- *G. rosmarinifolia*
1'-4'-5'-6'-7'- *Hakea* (p. 138) 1'-9'-14'-15-16-17- *H. sericea*

WILLIS (233)
Key to families, vol. 2, p. 1.
1-4-6-7-12-138(F)-144-145-146-147-148-149- Proteaceae

Key to genera, p. 32.
1- *Banksia* 1-2-4- *B. marginata*
1-2-3-4- *Persoonia* 1-2-4- *P. juniperina*
1-2-3-4- *Conospermum* 1- *C. mitchellii*
1-2-3-5- *Adenanthos* - *A. terminalis*
1-2-6-7- *Hakea* 1-4-8-9-10- *H. sericea*
1-2-6-7-8-9- *Grevillea* 1-2-13-14-19- *G. rosmarinifolia*

Plant Families

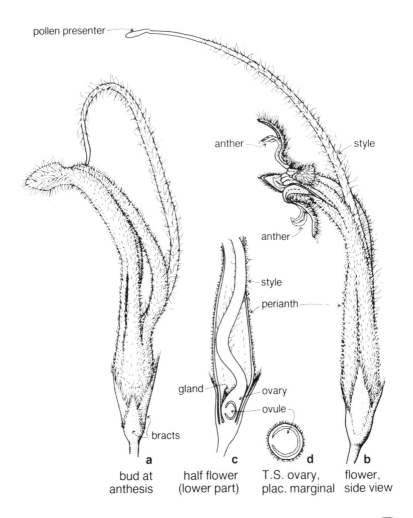

a bud at anthesis
c half flower (lower part)
d T.S. ovary, plac. marginal
b flower, side view

Fig. 26 *Adenanthos terminalis* (gland flower) Proteaceae   P⌢(4) A4 G1

Spreading shrub to 1 m. Branches slender, erect, young branches and leaves pubescent. Leaves crowded and overlapping, 0.5–1 cm long, digitately divided into 3–5 terete lobes. Flowers pale yellow, terminal, sessile and 1–3 per cluster. Four small nectary glands surround the ovary and alternate with the tepals. Fruit a pubescent nut. Sandy heaths of western Vic., south-eastern SA and Kangaroo I. Uncommon. Flowering in spring. (a–c ×5, d ×10)

PROTEACEAE

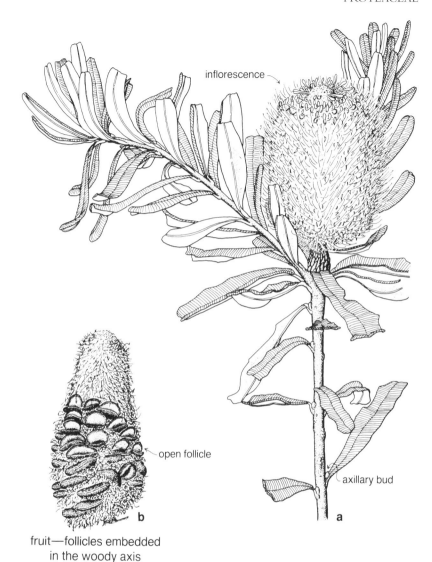

Fig. 27 *Banksia marginata* (silver banksia) Proteaceae

Habit variable, low stunted shrub or small tree. Young shoots bronze in colour due to T-shaped hairs. Leaves shortly stalked, leathery, 2–8 cm long with truncate apices, margins sometimes shortly toothed, underside white-tomentose, veins conspicuous. Flowers small, cream to yellow in a terminal spike-like inflorescence 4–10 cm long. Four small nectary glands alternate with the tepals at the base of the ovary. Fruit a follicle. Widespread in Vic., SA, Tas., and NSW. Flowering autumn to early spring. (a–b ×0.7)

Plant Families

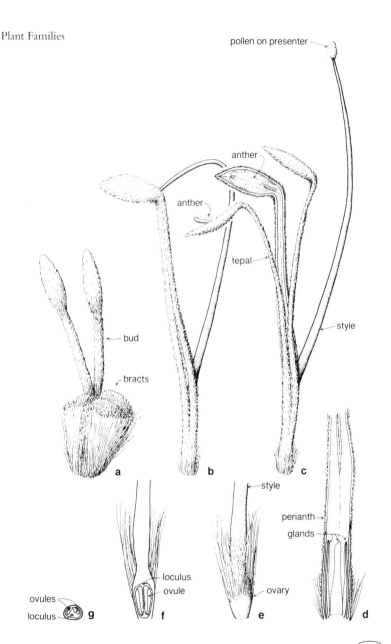

Fig. 28 *Banksia marginata* (silver banksia) Proteaceae   P⌢(4) A4 G1̲
a two flower buds with subtending bracts; **b** flower bud at anthesis, side view; **c** flower, side view; **d** base of perianth, internal view; **e** ovary and base of style, side view; **f** as in (e), but L.S.; **g** T.S. ovary; placentation marginal (a–c ×5, d–g ×10)

PROTEACEAE

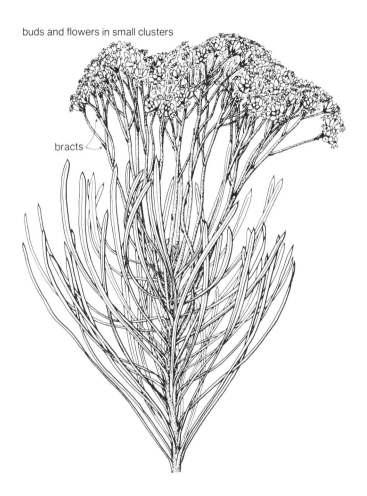

Fig. 29 *Conospermum mitchellii* (Victorian smoke-bush) Proteaceae
Spreading, often multi-stemmed shrub 1–2 m tall. Leaves linear 5–15 cm long, curving upwards. Inflorescence a corymbose panicle. Buds bluish-grey. Flowers white, each subtended by a dark bract. The lobes of adjacent anthers are at first united, and only 4 of the 8 lobes are fertile; the other 4 lobes are reduced to small appendages. The front stamen has 2 functional lobes, and the 2 lateral stamens have 1 functional lobe each. Fruit a pubescent nut. On sandy heaths at Anglesea in western Vic., and south-eastern SA. Flowering in spring. There are more than 40 species of *Conospermum* in WA, about 10 in eastern states. (×0.7)

Plant Families

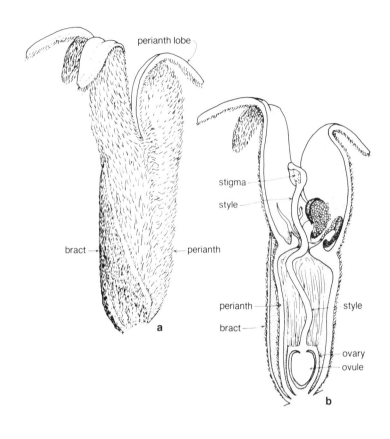

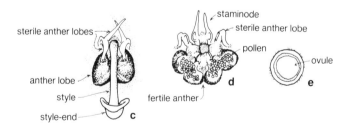

Fig. 30 *Conospermum mitchellii* (Victorian smoke-bush) Proteaceae   P(4) A4 G1
a flower and bract, side view; **b** half flower; **c** stamens and style, from above; **d** stamens dehisced, from above; **e** T.S. ovary, placentation apical (a–e ×12)

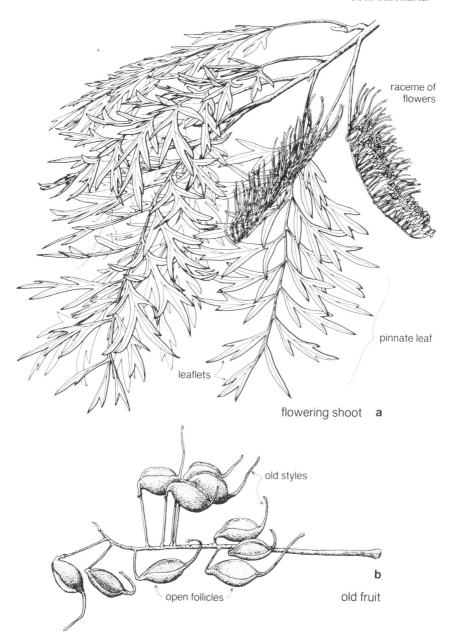

Fig. 31 *Grevillea robusta* (silky oak) Proteaceae (a ×0.3, b ×0.7)

Plant Families

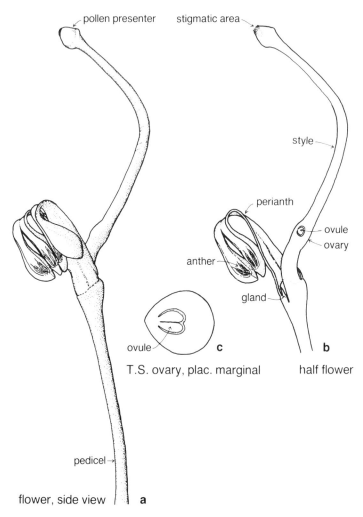

Fig. 32 *Grevillea robusta* (silky oak) Proteaceae   P(4) A4 G1
Tall forest tree to 30 m. Leaves pinnate, divided into 10–25 leaflets which are further divided or lobed. Flowers orange, borne in one-sided racemes. The ovary is shortly stalked. Rainforests and along creeks in Qld and northern NSW. Extensively grown in southern Australia in parks, gardens and as a street tree. In Vic., flowering early summer. (a–b ×3, c ×12)

PROTEACEAE

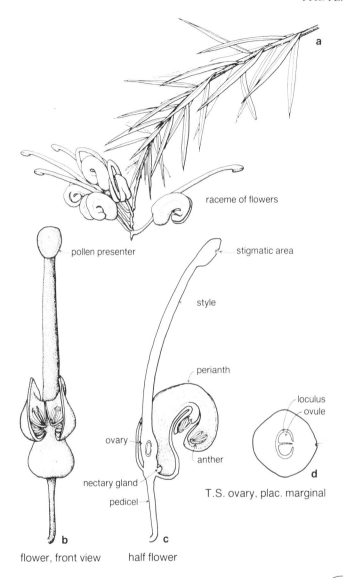

Fig. 33 *Grevillea rosmarinifolia* (rosemary grevillea) Proteaceae P(4) A4 G1
Spreading shrub to about 2 m high. Leaves linear, sharp-pointed, 2–4 cm long, often with recurved margins. Flowers red, pink or cream, often red and cream, in short racemes. Perianth glabrous externally. Fruit a follicle, style persistent in fruit. Widespread but scattered, mostly on dry rocky sites, Vic. and NSW. Extensively cultivated. Peak flowering time early spring. See also Pl. 2a–b. (a ×1.2, b–c ×3, d ×12)

83

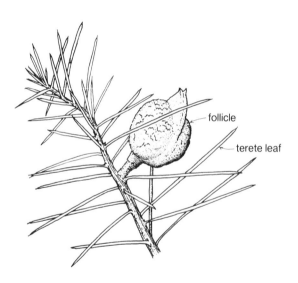

Fig. 34 *Hakea decurrens* (bushy needlewood) Proteaceae
Variable bushy shrub. Leaves terete, pungent, about 2–6 cm long, grooved below at the base. Flowers small, white or sometimes pink, in axillary clusters of 4–6. Fruit about 2–4 cm long, sometimes almost round, woody with a rough surface, a beak at the apex of each valve. Widespread, locally common in drier forests especially those with a heathy understorey. Vic., NSW, some Bass Strait islands. Flowering winter to spring. (×0.7)

PROTEACEAE

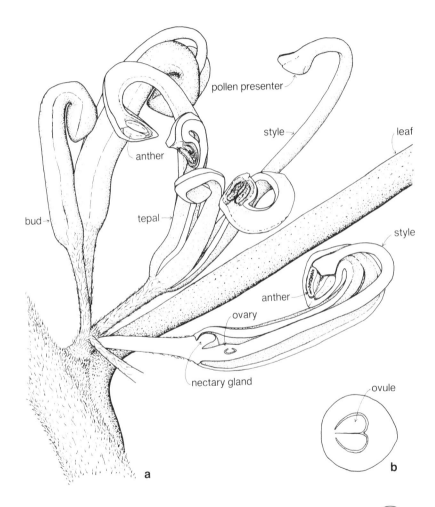

Fig. 35 *Hakea decurrens* (bushy needlewood) Proteaceae   P4 A4 G1
a flower cluster in the axil of a leaf—the lowest flower is shown in section; **b** T.S. ovary, placentation marginal. (a ×8, b ×25)

85

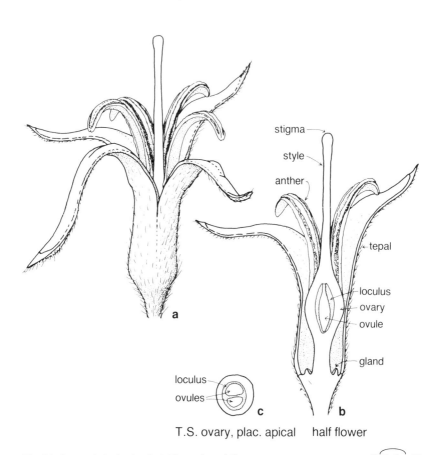

Fig. 36 *Persoonia juniperina* (prickly geebung) Proteaceae P(4) A4 G1

Bushy shrub, usually about 1 m high. Leaves linear, sessile, sharp-pointed, about 3 cm long, channelled on upper surface. Flowers yellow, solitary, shortly stalked and axillary. The ovary is shortly stalked with small nectary glands at the base. Fruit a drupe, blue-black at maturity. Widespread in heathlands and drier forests of Vic., NSW, Tas. and SA, especially near the coast. Flowering in summer. (a–c ×9)

LORANTHACEAE mistletoes

Most members of the Loranthaceae, and the closely related Santalaceae (sandalwoods) and Viscaceae, are root or stem parasites. The European mistletoe, *Viscum album*, which figures extensively in mythology and legends, belongs to Viscaceae. The type genus of Loranthaceae, *Loranthus*, formerly of about 600 species, does not occur in Australia and those Australian species that were believed to belong to *Loranthus* are now assigned to *Amyema* and some other genera.

Most members of Loranthaceae are stem parasites, and some are found on several hosts, including eucalypts, wattles, she-oaks and introduced trees such as oaks, birches, *Liquidambar*, planes and fruit trees. Others are specific to one host genus only. The Western Australian Christmas tree, *Nuytsia*, is a root parasite.

The seeds of Loranthaceae are surrounded by a sticky layer rich in glucose and are an important food source for birds, especially the mistletoe bird. The seeds pass through the bird's gut in 3–12 minutes and, as the bird twists when defecating, the seeds drop on the branch. The sticky layer is still present and the seed adheres to the branch, where it germinates immediately (Pl. 4 b). The parasite is attached to the host stem by one or more structures called haustoria, through which the respective vascular systems join up.

Amyema (Pl. 4a) is the largest genus of mistletoes in southern Australia. The bisexual flowers are borne in axillary umbel-like groups of 2 or 3. Each flower is usually subtended by a bract. The calyx is reduced to a rim of tissue on top of the inferior ovary. Petals are 4–6 in 1 whorl, free or united into a tube, which is often split down one side. Stamens are equal in number to the petals and epipetalous with basifixed anthers. The ovary structure is obscure and the ovules are undifferentiated. The fruit is berry-like and contains 1 seed. The stems are rather brittle, usually with thick, simple, opposite leaves, with parallel venation. In many cases the foliage of the parasite is visually similar to that of the host.

For descriptions and keys to Australian species, see *Flora of Australia* vol. 22 (bibliography no. 94).

Plant Families

CHENOPODIACEAE saltbushes and samphires

Members of the Chenopodiaceae are distributed worldwide, except for the tropics. They are common in saline habitats and arid environments. There are more than 100 genera, 32 of which occur in Australia; of these 28 are native and four are introduced.

A few species are cultivated for food, for example varieties of *Beta* (beetroot and sugar beet), and *Spinacia* (spinach). The fleshy fruits of *Enchylaena* and *Rhagodia* are edible and were utilised by Aboriginal people, and the seeds of *Atriplex* and *Tecticornia* were made into flour.

Some *Atriplex* spp. (saltbushes), particularly *A. nummularia* (old man saltbush), and *Maireana* spp. (bluebushes) provide fodder in the pastoral areas of the arid zone. A number of members of the family are common weeds of waste ground, roadsides and cultivation, e.g., *Chenopodium album* (fat hen).

Members of the family are illustrated in Fig. 37, and Plates 5 and 6d–e. For descriptions and keys to Australian species see *Flora of Australia* vol. 4 (bibliography no. 94).

FLORAL STRUCTURE

Flowers	Inconspicuous, actinomorphic. Solitary or few together (Pl. 6d) or arising in dense clusters (Pl. 5a), or as in the samphires, immersed in the stem tissue (Fig. 37b; Pl. 5d). Bisexual or unisexual. If the flowers are unisexual, the plants may be monoecious or dioecious.
Calyx	Sepals 5, usually united at the base, usually persistent and enlarging in fruit. Because petals are absent, the sepals are usually referred to as the perianth. In *Atriplex*, the perianth is absent and replaced by a pair of bracteoles (Pl. 5a).
Corolla	Petals absent.
Androecium	Stamens 1–5, opposite the perianth parts (Pl. 6d).
Gynoecium	Carpels 2, united. Ovary superior, with 1 loculus. The style bears 2 (sometimes 3) stigmatic branches (Fig. 37e–f; Pl. 5a).
Fruit	Usually a berry or nut. The perianth is persistent and in many genera may enlarge in fruit, often developing appendages such as wings (*Maireana*, Pl. 5c), spines (*Sclerolaena*), or tubercles (*Malococera*). The fruiting perianth is succulent in *Enchylaena* (Pl. 5e). In *Atriplex*, the bracteoles enlarge (Pl. 5a) and often develop appendages, or become bladdery. The fruits of samphires are found in the dried-up parts of the stems where flowering occurred.

CHENOPODIACEAE

The family is comprised of herbs and shrubs. The leaves are generally small, often succulent, and crowded on the branches. Young plants are often covered with tiny bladder-like hairs, which later deflate to give the plant a greyish or mealy appearance (Pl. 6d, e). The flower structure is relatively uniform, and fruit characters are used extensively in the identification process.

SPOTTING CHARACTERS

Small shrubs or herbs mostly with small fleshy leaves, often mealy. Flowers small and inconspicuous, with a single perianth whorl. Stigmas usually 2.

ROUTES THROUGH KEYS

Routes through several keys are given here for the genus *Sarcocornia* (previously known as *Salicornia*, Fig. 37). Texts are cited by the author's name followed by the bibliography number.

BEADLE (18)
Key to families unsatisfactory for this genus.

BLACK (24)
Key to families, p. 23 (see note about Black's key to families, in the section on Proteaceae). Alternative leads in a couplet are here designated A1, A2, A3; B1, B2, etc.
A2-B2-C2- subclass 1(p. 24) -A2-B4-D1-E1- Chenopodiaceae

Key to genera, p. 286.
A2-H2- *Salicornia* - *S. australis* (= *S. quinqueflora*)

GEORGE/ORCHARD (94)
The key does not cope well with the variation in the number of perianth lobes. The second route (perianth 4-partite / segments 4–6) requires accepting a later lead stating that the *leaves* are 'covered with mealy scales'.

Key to families, vol. 1, (1st edn) p. 113.
1-2-Dicotyledons-3-10-572-647-652-653-667-('perianth segments 3')
   -670- Chenopodiaceae.
1-2-Dicotyledons-3-10-572-647-652-653-667-('perianth segments 4')
   -674-675-701-702-703-704- Chenopodiaceae

Key to families, vol. 1, (2nd edn) p. 521.
1-2-3-10-693-777-782-783-802-('perianth segments 1–3')-803-804-807
   -808-Chenopodiaceae
1-2-3-10-693-777-782-783-802-('perianth segments 4–6')
   -812-814-815-846-847-848-849-Chenopodiaceae

Plant Families

JESSOP AND TOELKEN (127)
Key to families vol. 1, p. 65. The second lead of each pair is here designated by ' following the number. The key to families is scarcely satisfactory for this genus as, at couplet 48, you need to assume that *leaves* are alternate, or if opposite, then flat.
1'-3'-4- Dicotyledonae (to p. 66) 1'-2'-3'-8'-9-10'-32'-33'-43'-47-48'-49'-52- Chenopodiaceae

Key to genera, vol. 1, p. 236.
1'-25- *Sarcocornia*

WILLIS (233)
Key to families, vol. 2, p. 1.
1-4-6-7-12-13-178(H)-179-180-181-182- Chenopodiaceae

Key to genera, p. 80.
1-12-13- *Salicornia* -(p. 110)- *S. quinqueflora*

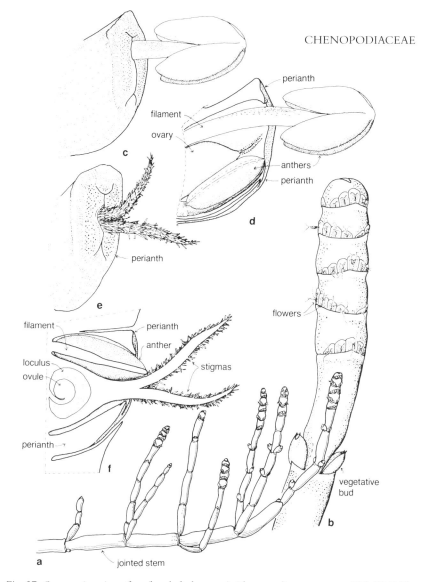

Fig. 37 *Sarcocornia quinqueflora* (beaded glasswort) Chenopodiaceae P(4) A2 G($\underline{2}$)
a flowering branch; **b** flowering stem apex; **c** flower, side view, with stamen protruding from the fleshy perianth; **d** same flower, sectioned—the small ovary may be infertile; **e** flower, side view, with stigmas protruding; **f** same flower, sectioned—placentation basal (a ×0.6, b ×3, c–f ×24)

Decumbent or shortly erect or tufted perennial with apparently leafless stems made up of fleshy cylindrical segments. Inflorescence spike-like with flowers in small groups immersed in the stem tissue towards the tips of branchlets. Perianth segments 3 or 4, succulent, with the upper lobe overlapping the lateral ones. Locally common in saltmarsh communities of all states except NT, mainly coastal, but occasionally fringing saline lakes inland. Flowering mainly in autumn. See also Pl. 5d.

RANUNCULACEAE buttercup family

This large worldwide family is mainly of the temperate zones. The name of the type genus *Ranunculus* (buttercup, Pl. 1f) comes from a Latin word meaning a little frog or tadpole and, as it might suggest, many buttercups are marsh plants. Most members of the family are herbaceous but some, including *Clematis*, are climbers.

Clematis microphylla (small-leaved clematis, Fig. 38) is a woody climber, common in temperate parts of Australia. The flowers have no petals, are usually unisexual and the inflorescence is a panicle. The female flowers have 4 creamy, oblong sepals and alternating with them are 4 staminodes with long filaments and non-functional anthers. Each of the many free carpels has a long recurved style. The male flower has similar sepals and many stamens. At the fruiting stage the female flowers become a cluster of achenes, each of which bears a long, conspicuous, plumose awn derived from the style.

RANUNCULACEAE

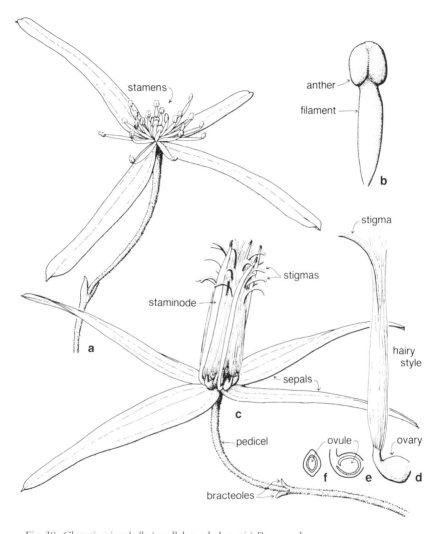

Fig. 38 *Clematis microphylla* (small-leaved clematis) Ranunculaceae
K4 C0 A∞ G0 (♂)   K4 C0 A staminodes G∞ (♀)
a male flower; **b** stamen; **c** female flower; **d** carpel, side view; **e** L.S. ovary, placentation apical; **f** T.S. ovary (a ×3, b ×12, c ×3, d–f ×7)

93

LAURACEAE laurel family

This family is widespread and particularly well developed in tropical rainforests. The family name is derived from the genus *Laurus* (laurels) and in ancient times the leaves of *L. nobilis* (bay laurel) were used as the symbol of victory. The aromatic leaves are now used for flavouring soups and meat dishes. Other important cultivated members of the family are *Persea americana* (avocado), *Cinnamomum camphora* (camphor laurel), and *C. zeylanicum*, the bark of which yields the spice cinnamon. Genera growing in northern rainforests include valuable timber trees.

A marked contrast in form is found in the genus *Cassytha* (dodder-laurel), widespread in southern Australia. The members of this genus are parasitic perennials, with tough twining stems. They are attached to the host plants by small sucker-like pads called haustoria, through which the vascular system of the parasite is linked to that of the host (Pl. 4f).

The leaves are reduced to minute scales. The flowers are small, regular and borne in short racemes or spikes. There are 6 white to yellow perianth parts in 2 whorls of 3, and 9 stamens in 3 whorls of 3. Note that flowers with parts in threes are unusual in the dicots. The ovary is superior and unilocular with 1 pendulous ovule. The fruit is a berry or a drupe.

Cassytha glabella (slender dodder-laurel) is common in heathlands, and *C. melantha* (coarse dodder-laurel, Pl. 4f), which has much thicker stems, is found on trees adjoining heathland and in forests.

DROSERACEAE sundews

The sundews have a world wide distribution and there are about 100 species of the largest genus, *Drosera*, in Australia (Pl. 4e). Droseras are annual or perennial herbs, commonly growing each season from a tuberous structure which lies dormant underground during the summer. They vary in habit from rosettes to small erect plants or straggly climbers with stems up to 60 cm long. The leaves may be more or less round, or strap-shaped or spoon-shaped, or sometimes forked, and the upper surfaces are covered with sensitive glandular hairs. Small animals, such as ants or mosquitoes, are trapped by a sticky substance exuded by the hairs, which then fold over to prevent escape. Other hairs secrete enzymes that digest the prey. By this means the plants are believed to obtain nitrogenous products that supplement their photosynthetic nutrition.

The flowers of *Drosera* are regular and bisexual, mostly with 5 sepals, 5 white or coloured petals, and numerous stamens. The ovary is superior, unilocular with 2–5 styles and 3–many basally attached ovules. The fruit is a capsule.

For further reading, see numbers 80, 94 and 149 in the bibliography.

Plant Families

TREMANDRACEAE

This small family of three genera is confined to southern Australia. Two of the genera, including *Tremandra* from which the family name is derived, are confined to south-western Australia but the largest genus, *Tetratheca*, is found in all states but not the Northern Territory.

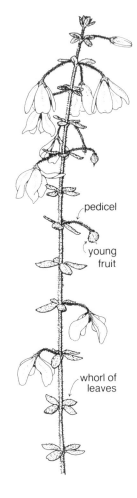

Fig. 39 *Tetratheca ciliata* (pink-bells) Tremendraceae   (×0.7)

TREMANDRACEAE

The flowers of *Tetratheca* in the eastern states have 4 small sepals and the same number of larger petals. There are 8 stamens with short filaments and large 4-celled, erect, usually dark-coloured anthers. The anther lobes contract at the top to form a short tube and the pollen is shed through a single apical pore. The ovary is superior with 2 loculi, which in most species contain 1–2 usually pendulous ovules. The flowers are borne singly, or a few together, in the axils of the leaves, and the fruit is a capsule. Most of the Western Australian species and one from Kangaroo I. are 5-partite but have a similar ovary structure.

Tetratheca ciliata (pink-bells, Figs 39, 40; Pl. 1d) is a variable small shrub, common in heaths and forests across southern Vic. and adjacent parts of SA and NSW. It is occasionally seen in cultivation. The leaves are extremely variable in size and may be alternate or opposite, or more often in whorls of 3 or 4. The vivid purplish-pink flowers are conspicuous during the spring and early summer.

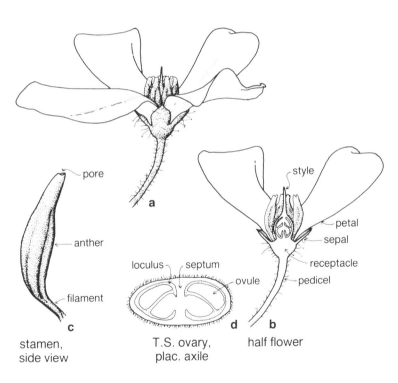

Fig. 40 *Tetratheca ciliata* (pink-bells) Tremandraceae   K4 C4 A8 G(2)
(see also Pl. 1d; a–b ×3, c–d ×12)

PITTOSPORACEAE pittosporum family

Most of the nine genera belonging to this family are restricted to Australia. *Pittosporum*, from which the family name is derived, is the largest genus with about 150 species, and it extends into South-east Asia and across to Africa.

The berries of *Billardiera scandens* (apple-berry, Fig. 42) are edible and were used by Aboriginal people. *Pittosporum angustifolium* (weeping pittosporum, native willow, Fig. 43), *P. undulatum* (sweet pittosporum, Fig. 44) and *Bursaria spinosa* (sweet bursaria) have been used as hedge plants and as ornamentals. *Hymenosporum flavum* (native frangipani) and the climbers *Sollya heterophylla* (bluebell creeper), *Billardiera bignoniacea* (orange bell-climber, Fig. 41) and other *Billardiera* species are common in native gardens. *Pittosporum undulatum*, which is native to the wetter forests of eastern Australia, has become an environmental weed in some southern coastal and forest regions.

FLORAL STRUCTURE

Flowers	Actinomorphic, bisexual.
Calyx	Sepals 5, free or united at the base, falling early in some *Pittosporum* species.
Corolla	Petals 5, sometimes clawed, as in *P. undulatum* (Fig. 44), usually free but sometimes wholly or partially coherent.
Androecium	Stamens 5, free.
Gynoecium	Carpels 2–5, united, ovary superior with 2–5 parietal or axile placentas. Ovary with 1–5 loculi. A unilocular ovary is sometimes incompletely divided by the inward growth of parietal placentas (Fig. 44).
Fruit	A capsule or berry.

The family consists of trees, shrubs and climbers. The leaves are simple, usually alternate, sometimes whorled. There are no stipules.

SPOTTING CHARACTERS

Flowers actinomorphic, 5-partite. Ovary superior, often of 2 united carpels.

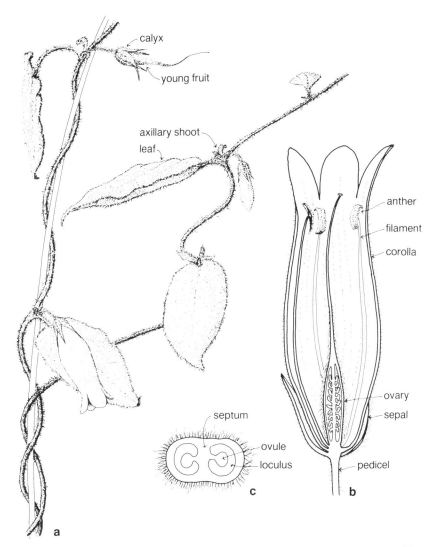

Fig. 41 *Billardiera bignoniacea* (orange bell-climber) Pittosporaceae   K5 C(5) A5 G(2)
a part of two twining leafy stems with axillary flowers; **b** half flower; **c** T.S. ovary, placentation axile (a ×1.2, b ×4, c ×16)

Twining creeper, stems becoming glabrous with age. Leaves hairy, scattered, ovate or cordate, 2–4 cm long. Flowers pendulous on long slender stalks, 1–3 in axils of the leaves. Corolla greenish at the base, orange to salmon-coloured above. Fruit a pubescent capsule. Shady damp places in the Grampians, Vic., and in SA on Kangaroo I. and Mt Lofty Ranges. Flowering mainly in summer. Previously known as *Marianthus bignoniaceus*.

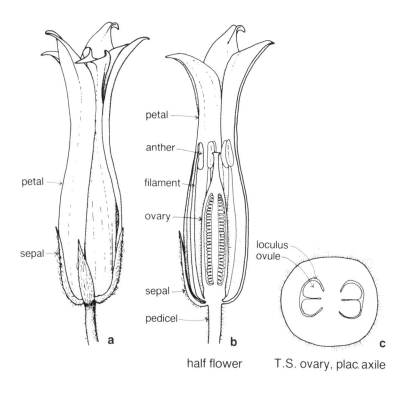

Fig. 42 *Billardiera scandens* (common apple-berry) Pittosporaceae   K5 C5 A5 G($\underline{2}$)
Twining creeper. Leaves lanceolate to linear, usually with undulate margins, 1.5–5 cm long. Flowers pale yellow, pendulous, solitary and axillary. Corolla appearing tubular but petals free. Fruit an elongated greenish berry. Widespread in forests of Vic., SA, Tas., NSW and Qld. Flowering spring to early summer. (a–b ×4, c ×12)

PITTOSPORACEAE

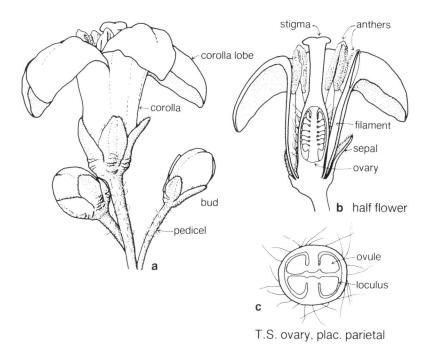

Fig. 43 *Pittosporum angustifolium* (weeping pittosporum) Pittosporaceae

K5 C(5) A5 G(2)

Shrub or small tree to 7 m high. Branches pendulous. Leaves glabrous, narrow-linear, 4–10 cm long. Flowers yellow, solitary or in small axillary cymose clusters. Petals often virtually united when young, becoming free with age. Fruit an ovoid orange capsule 1–2 cm long, seeds red and sticky. Dry inland areas of all states except Tas. Flowering winter to spring. (a–b ×5, c ×10)

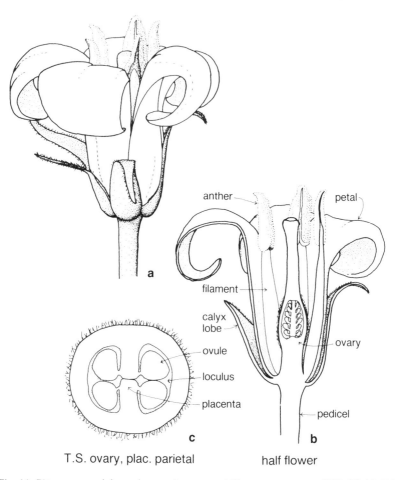

T.S. ovary, plac. parietal   half flower

Fig. 44 *Pittosporum undulatum* (sweet pittosporum) Pittosporaceae   K(5) C5 A5 G(2)
Bushy tree to 15 m high. Leaves glabrous, lanceolate, 6–12 cm long, sometimes appearing whorled near the ends of branches, margins undulate. Flowers cream, fragrant, in leafy clusters. In cultivation, plants with only female flowers quite common. Fruit a globular orange capsule, seeds brown and sticky. In Vic. originally restricted to the wet gullies of the east, but now widely planted in hedges and parks, and a bad weed in some areas, particularly near-coastal. Also in NSW, Qld, rare in north-western Tas. (a–b ×5, c ×20)

ROSACEAE rose family

The Rosaceae is a large cosmopolitan family, well developed in the northern hemisphere, but with only a few genera native to Australia. The family name is derived from the genus *Rosa* (rose).

Economically the family is very important, as many members produce edible fruit. Examples include *Malus* (apple), *Prunus* (almonds, plums and other stone fruits), *Rubus* (blackberry, loganberry, raspberry), *Pyrus* (pear), *Chaenomeles* (quince) and *Fragaria* (strawberry). Familiar ornamental genera include *Crataegus*, *Rosa*, *Cotoneaster* and *Spiraea*. *Crataegus monogyna*, *C. oxyacanthoides* (hawthorn) and their hybrids, *Rubus fruticosus* (blackberry) and *Rosa rubiginosa* (sweet briar), all of European origin, are declared noxious weeds in Victoria and have become very troublesome in south-eastern Australia. Native species of *Acaena*, including *A. novae-zelandiae* (bidgee-widgee), and *A. echinata* (sheep's burr, Fig. 45) produce burrs that adhere to clothing and the coats of animals. Of the five members of *Rubus* native to Australia, *R. parvifolius* (small-leaf bramble or native raspberry) is widely spread in south-eastern Australian forests.

Taxonomically the family is a very difficult one. It has been divided into five subfamilies and numerous tribes but there is little consensus on the composition of these groups. In the *Flora of New South Wales* (bibliography no. 104), those species occurring in that state traditionally included in Rosaceae have been treated in three families, the Rosaceae (in a narrower sense), Malaceae and Amygdalaceae.

Members of the family are illustrated in Figures 45–7.

FLORAL STRUCTURE

Flowers	Actinomorphic, usually bisexual. Floral tube usually present (and flowers perigynous), sometimes epigynous or hypogynous.
Calyx	Sepals 4–5, imbricate, usually free. An epicalyx is often present and then its units alternate with the sepals.
Corolla	Petals 4–5, imbricate, free, sometimes absent, sometimes falling early. In many cultivars, such as roses, the number of petals has increased due to stamens becoming petaloid.
Androecium	Stamens usually numerous, rarely 1, 2 or few.
Gynoecium	Carpels 1 or 2 to many, often free. Ovary superior or inferior.
Fruit	The type of fruit is related to the structure of the floral tube, and to whether the flower was perigynous, epigynous or hypogynous. The following describes the development of some familiar fruits.

Flower perigynous

In *Acaena* (Fig. 45) and *Prunus* (Fig. 47), a single carpel is attached to the base of the floral tube. The fruit is either an achene enclosed in the dried floral tube (*Acaena*) or a drupe (*Prunus*). The fruit of the almond is also a drupe, as it

Plant Families

develops in the same way as a plum. The shell of the almond is equivalent to the stone of the plum, as the fleshy outer husk of the almond dries up and falls off. The edible part is the seed. In roses, many free carpels are enclosed in the urn-shaped floral tube. Each carpel contains two ovules. The fruit is a fleshy 'hip' derived from the floral tube and containing many hairy achenes.

Flower epigynous

In apples, pears and quinces, the carpels are fused to the inner wall of the base of the floral tube and so the ovary is inferior (Fig. 46). The fruit is a pome, formed by the growth of the tube wall, which becomes fleshy and encloses the fleshy ovary wall. Thus the 'flesh' of these fruits has mostly developed from the floral tube.

Flower hypogynous

In raspberries and blackberries, the receptacle is more or less flat and the free carpels are borne on a fleshy stalk, the gynophore. Each carpel develops into a small drupe. In the strawberry, the green receptacle is swollen and fleshy with the carpels attached to the outside. At maturity the receptacle becomes soft and coloured, with the tiny achenes embedded in the surface.

Most members of the family are shrubs or perennial herbs and there are a few annuals and climbers. The leaves are usually alternate, simple or compound and nearly always have stipules, although these are sometimes very small. Thorns, which are modified branches, are found in *Crataegus* and *Prunus*, but the prickles of *Rosa* and *Rubus* are surface appendages.

SPOTTING CHARACTERS

Flowers actinomorphic, often perigynous with numerous stamens. Stipules often present.

ROUTES THROUGH KEYS

Routes through several keys are given here for the genus *Acaena* (Fig. 45). Texts are cited by author's name followed by the bibliography number.

BEADLE *et al.* (18)
Key to families, p. 99.
A- dicots -*A-*B-*C-*D-E-F-G-H-I- group 2(p. 101)
    *A-*B-*E-*F-G-*H-I-*J- Rosaceae

Key to genera, p. 258.
-A-*B- *Acaena* -A-*B-*C- *A. echinata*

ROSACEAE

BLACK (24)

Key to families, p. 23 (see note about Black's key to families, in the section on Proteaceae). Alternative leads in a couplet are here designated A1, A2, A3; B1, B2, etc.
A2-B2-C2- subclass 1(p. 24) -A3 (p. 26) -F2-H3-K2-L2- Rosaceae

Key to genera, p. 395.
A2-C2-D2- *Acaena*

GEORGE/ORCHARD (94)

The key to families in both editions of the *Flora of Australia* vol. 1 is unsatisfactory for the genus *Acaena*.

JESSOP AND TOELKEN (127)

Key to families, vol. 1, p. 65. The second lead of each pair is here designated by ' following the number. This key is scarcely satisfactory for the genus *Acaena*, as at couplet 33, it must be assumed that the number of stamens is equal to or less than the number of perianth segments. This will be true in some cases but stamen number is variable in this genus.
1'-3'-4- Dicotyledonae (to p. 66)
   1'-2'-3'-8'-9-10'-32'-33'-43'-47'-53'-61'-62- Rosaceae

Key to genera, p. 437.
1-2'-3'-4- *Acaena*

WALSH AND ENTWISLE (219)

Key to families, vol. 3, p. 5.
1-2-3-4-9-11-12-13-18-19- Group 8 -1-2-4-5- Rosaceae

Key to genera, p. 556.
1-2-3-4-5-6-7-9- *Acaena* (p. 564) 1-2-3- *A. echinata*

WILLIS (233)

Key to families, vol. 2, p. 1.
1-4-6-7-12-13-178(H)-179-180- Rosaceae

Key to genera, p. 201.
1-6-9-11- *Acaena* -1-2- *A. echinata*

105

Plant Families

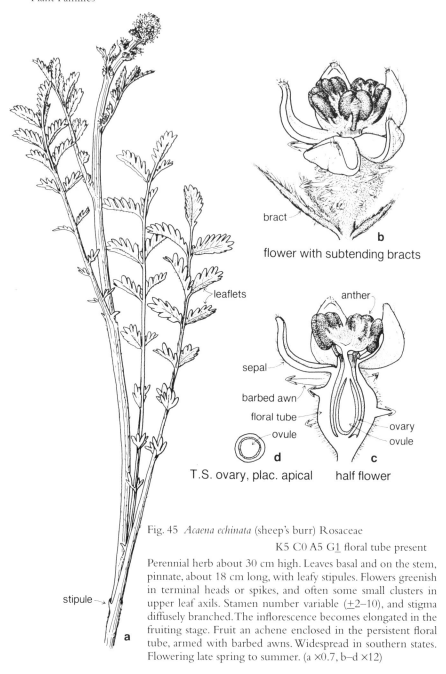

Fig. 45 *Acaena echinata* (sheep's burr) Rosaceae
K5 C0 A5 G<u>1</u> floral tube present

Perennial herb about 30 cm high. Leaves basal and on the stem, pinnate, about 18 cm long, with leafy stipules. Flowers greenish in terminal heads or spikes, and often some small clusters in upper leaf axils. Stamen number variable (±2–10), and stigma diffusely branched. The inflorescence becomes elongated in the fruiting stage. Fruit an achene enclosed in the persistent floral tube, armed with barbed awns. Widespread in southern states. Flowering late spring to summer. (a ×0.7, b–d ×12)

ROSACEAE

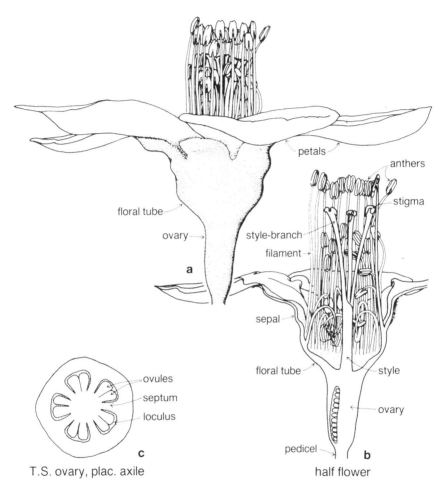

T.S. ovary, plac. axile   half flower

Fig. 46 *Chaenomeles speciosa* 'Nivalis' (japonica or flowering quince) Rosaceae

K5 C5 A∞ G($\overline{5}$) floral tube present

Shrub to about 3 m high, often thorny. Leaves oblong-ovate, up to 8 cm long, hairy underneath when young, margins serrate. Flowers white (pink or red in other forms), solitary. Male flowers, without ovaries, are not uncommon. Fruit fleshy, yellow-green. Widespread in cultivation. Origin, Japan. Flowering winter to spring. (a–b ×3, c ×9)

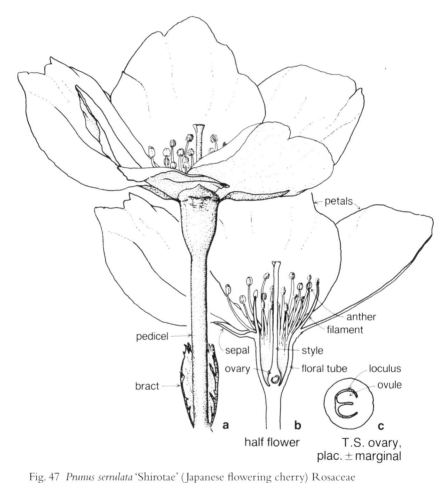

Fig. 47 *Prunus serrulata* 'Shirotae' (Japanese flowering cherry) Rosaceae

K5 C5 A∞ G1 floral tube present

Small ornamental tree. Leaves ovate, acuminate, up to 10 cm long, margins serrate. Flowers white, each subtended by a fringed bract, in clusters of 3–5. One of many cultivars derived from *P. serrulata*. Origin, Japan. Flowering in spring. (a–b ×2, c ×9)

The order FABALES legumes
(LEGUMINALES)

The status of members of this order varies with the reference consulted. The *Flora of Australia* (vol. 1, 1981) recognises three families: Mimosaceae, Caesalpiniaceae and Fabaceae (Papilionaceae). Some authors unite these into a single family, Leguminosae, with three subfamilies, Mimosoideae, Caesalpinioideae and Faboideae (sometimes called Papilionoideae or Papilionatae). The legume is the fruit type characteristic of the majority of members of the order.

MIMOSACEAE mimosa family, wattles

The family name is derived from the genus *Mimosa*, which has about 300 species, mostly in tropical America. The family is common throughout the tropics, subtropics and warm temperate regions of the world. Of the 17 Australian genera the largest by far is *Acacia* (wattle) with more than 950 species. *Acacia* is the only genus in this family native to Victoria, although *Paraserianthes lophantha* (cape wattle) is now naturalised there.

The wattles are conspicuous in natural woodlands and are widely planted in shelter belts and as ornamentals. Many are important commercially, as several species, particularly *Acacia melanoxylon* (blackwood), produce fine timber for furniture and the bark of many others is rich in tannins, which are used for tanning leather. Throughout Australia the wattles were important to Aboriginal people, who used the seeds for food and the bark, wood and gum for many purposes.

Several useful field guides to the wattles are available—see bibliography numbers 5, 185 and 226. For descriptions and keys to all Australian spp. see *Flora of Australia* vols 11A, 11B (bibliography no. 94, electronic version no. CD4), and for descriptions, line drawings and keys for species in south-eastern Australia see number 212. Other works on wattles include numbers 103, 175 and 205.

The following notes refer only to the genus *Acacia* which is illustrated in Figures 48–52 and Plate 5a–c.

FLORAL STRUCTURE

Flowers	Small, sessile, actinomorphic, usually bisexual (Figs 49, 50). Arranged in globular heads (Figs 48, 50–2; Pl. 7a–b) or cylindrical spikes (Fig. 49; Pl. 7c). The heads or spikes are usually pedunculate and arranged singly, in pairs or racemes in the axils of the leaves or phyllodes.
Calyx	Sepals 4–5, free or united.
Corolla	Petals 4–5, usually free, valvate in bud.
Androecium	Stamens numerous, free. The most conspicuous part of the flower.
Gynoecium	Carpel 1. Ovary superior. Placentation marginal.
Fruit	A legume, often rather elongated, opening at maturity along both sutures. Sometimes the orientation of the seeds within the legume, and the form of the funicle are useful diagnostic features.

Members of the genus are trees or shrubs, the latter sometimes prostrate. Some species have bipinnate leaves (Fig. 52) but in most the foliage consists of entire, leaf-like structures called **phyllodes** (Figs 48–51; Pl. 7a–c), which are generally regarded as modified petioles. The phyllodes may be flat, broad to narrow, or terete. In the phyllodinous species the seedling leaves are bipinnate and the phyllodes form the adult foliage. In many wattles, and other members of the order Fabales, a small swelling may occur at the base of the phyllodes or bipinnate leaves or leaflets. Such a swelling is called a **pulvinus** (Figs 51, 52) and it is concerned with the movement of the leaf or phyllode in response to certain stimuli such as high wind or the onset of night.

Small glands (Figs 51, 52) often occur on the margins of the phyllodes, or on the petiole or rachis of the bipinnate leaves. These are sometimes referred to as extra-floral nectaries. Presence or absence of glands, and their number and position, are characters used in identification. Stipules are occasionally present (Pl. 7b) and in some species such as *A. paradoxa* (hedge wattle) they appear as spines. The venation of the phyllodes is another important character; one or a number of parallel veins may be evident (Pl. 7b–c).

SPOTTING CHARACTERS

Woody plants, flowers small and yellow in fluffy heads or spikes. Fruit a legume.

Plant Families

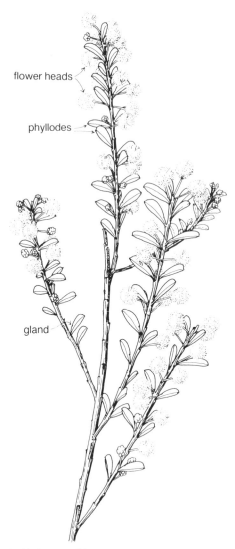

Fig. 48 *Acacia acinacea* (gold-dust wattle) Mimosaceae
Bushy shrub to 1.5 m. Branches erect or arching, ribbed, and marked with prominent phyllode bases. Phyllodes oblong, 0.5–2 cm long, glabrous, with one main vein, obtuse, mucronate. Glands, one often at the phyllode apex and one about midway along the upper margin. Flower heads globular, bright yellow, 1–4 per axil. Peduncles 1.5 cm long. Legume often curved, 2–5 cm long. Dry inland woodlands of Vic., NSW and SA. Flowering late winter to spring. See also Pl. 7. (×0.7)

MIMOSACEAE

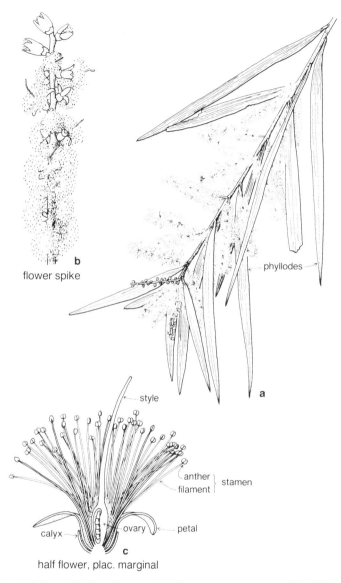

Fig. 49 *Acacia floribunda* (white sallow wattle) Mimosaceae K(4) C4 A∞ G1
Shrub or small tree to 7 m. Phyllodes lanceolate, straight or curved, 6–15 cm long, with fine veins. Flowers pale yellow, in loose spikes about 5–6 cm long, the finely pubescent rachis visible between the flowers. Spikes 1 or 2 together, axillary. Forests of eastern Vic., NSW and Qld. Widely cultivated. Flowering in spring. See also Pl. 7. (a ×0.6, b ×3, c ×12)

Plant Families

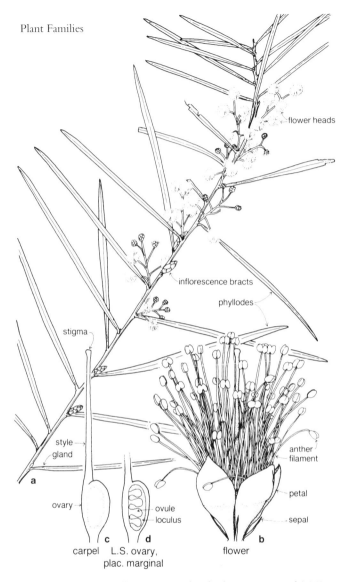

Fig. 50 *Acacia iteaphylla* (winter wattle, Flinders Range wattle) Mimosaceae

K5 C5 A∞ G1

Tall shrub to 4 m, young branchlets angular. Phyllodes linear 5–14 cm long, glaucous or green, with one main vein. Apex acute, mucronate. Gland small, near the base of upper margin. Flower heads globular, pale yellow, 8–12 borne in axillary racemes, enclosed at first in brown, ovate, deciduous bracts. Legume linear to 12 cm long. Endemic in southern SA, on hillsides and rocky valleys. Widely cultivated in southern Australia. Flowering in winter. See also Pl. 7. (a ×0.5, b–d ×20)

MIMOSACEAE

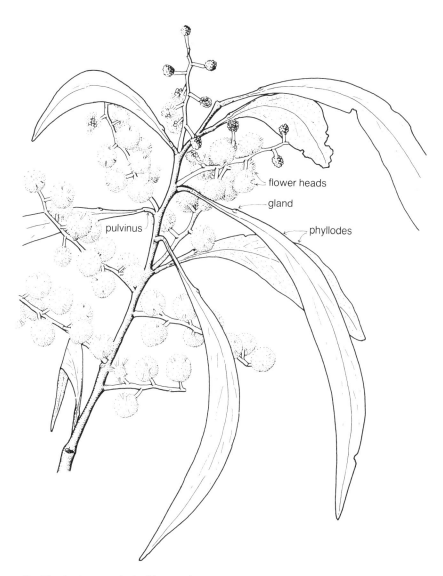

Fig. 51 *Acacia pycnantha* (golden wattle) Mimosaceae
Small tree to about 8 m. Phyllodes broad-lanceolate, falcate, 6–20 cm long, tough, midrib prominent, and lateral veins distinct. Gland conspicuous on upper margin. Flower heads globular, golden-yellow, fragrant, borne in axillary racemes or small panicles. Open forests of Vic., NSW and SA. Widely cultivated. Flowering in spring. See also Pl. 7. (×0.7)

Plant Families

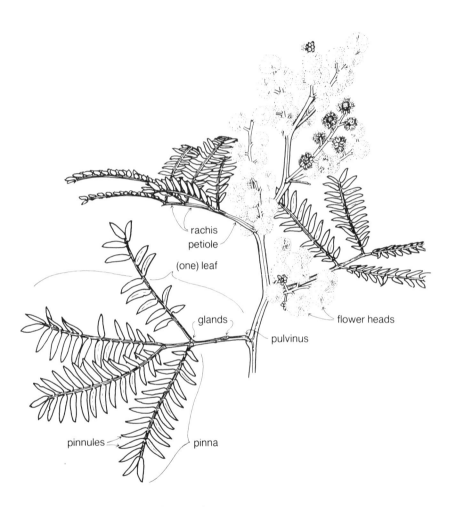

Fig. 52 *Acacia terminalis* (sunshine wattle) Mimosaceae
Shrub or small tree to 5 m. Branchlets angular. Leaves bipinnate, with 2–6 pairs of pinnae, each with 8–16 pairs of pinnules. Glands present on the petiole and at the bases of the pinnae. Flowers pale to golden yellow in globular heads. Heads in axillary racemes or small terminal panicles. Legume straight, up to 11 cm long. Open forests of eastern Vic. and NSW. Widely cultivated. Flowering late summer to autumn. Previously known as *A. botrycephala*. See also Pl. 7. (×0.7)

CAESALPINIACEAE cassias, sennas

The family takes its name from the genus *Caesalpinia*, which is confined to the tropics, with several species in northern Australia. It contains about 150 genera, distributed through the tropics and subtropics, with some members extending to the warmer deserts. One genus, *Senna*, is found throughout the centre of Australia, in all states except Tasmania.

Several genera are planted as ornamentals, including *Cassia*, *Senna*, the introduced *Gleditsia* (honey locust) and *Ceratonia* (carob bean). The senna of medicinal value is derived from the pods and leaves of *Senna* spp., and the fruits of *Cassia fistula* are also used medicinally. The fruits of *Tamarindus* (tamarind) have many culinary uses.

For descriptions and keys to all Australian species see *Flora of Australia* vol. 12 (bibliography no. 94) and for further reading see bibliography numbers 68, 180.

FLORAL STRUCTURE

Flowers	Usually zygomorphic, bisexual.
Calyx	Sepals 5, free.
Corolla	Petals 5, usually free, imbricate in bud with the posterior petal innermost.
Androecium	Stamens 10, but 3 often reduced to staminodes. In *Senna* (Figs 53, 54) the anthers open by terminal pores.
Gynoecium	Carpel 1. Ovary superior. Placentation marginal.
Fruit	A legume, often with partitions between the seeds.

Members of the family are trees or shrubs, some are woody climbers. Leaves are usually pinnate or bipinnate. Most sennas have pinnate leaves with no terminal leaflet.

SPOTTING CHARACTERS

Trees or shrubs. Leaves usually compound. Flowers zygomorphic, sometimes not markedly so, with a single carpel. Ovary superior. Fruit a legume.

Plant Families

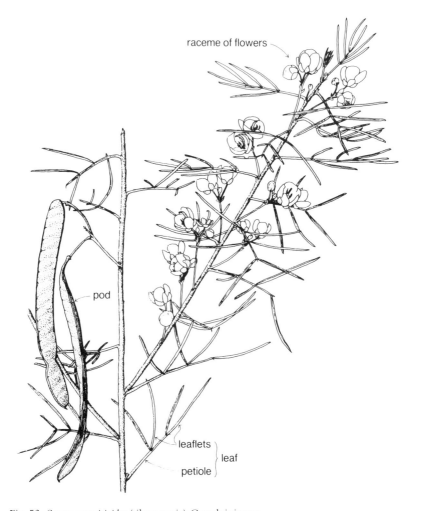

Fig. 53 *Senna artemisioides* (silver cassia) Caesalpiniaceae
Flowering branch, with two of the previous season's pods (×0.7)

CAESALPINIACEAE

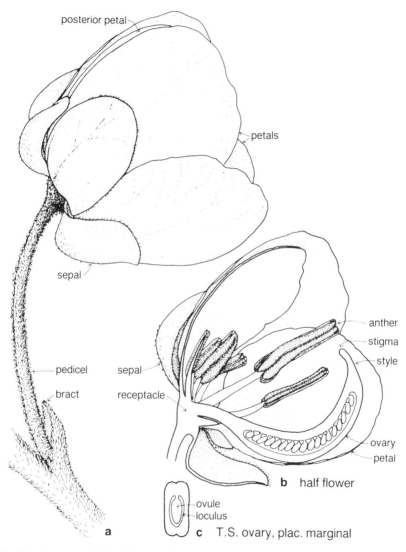

Fig. 54 *Senna artemisioides* (silver cassia)   K5 C5 A10 G1
Shrub to 2 m. Leaves compound, 3–6 cm long, leaflets linear. Flowers yellow, in axillary racemes. Legumes 8–10 cm long. Inland Australia. Widely cultivated in Vic. Flowering mainly in spring. Previously known as *Cassia artemisioides*. (a–b ×7, c ×12)

FABACEAE pea family

(PAPILIONACEAE)

This is a very large family mainly found in the temperate regions of the world, but extending to the tropics. The family name is derived from the genus *Faba*, which is the Latin word for a bean, but this generic name is no longer in use. The older name Papilionaceae is based on another Latin word, *papilio*, meaning a butterfly and applied because the shape of the flower is likened to that of a butterfly. The family is of great economic importance as many genera include common food plants and others are important for crops and pastures.

The seeds and pods of many species are important sources of food because they are rich in protein and minerals. These include *Pisum sativum* (garden pea), *Vicia faba* (broad bean), *Phaseolus vulgaris* (french bean), *Glycine max* (soybean), and *Arachis hypogea* (peanut). Species of *Trifolium* (clovers, Pl. 7d) and *Medicago* (lucerne, medic) are well-known pasture plants, and *Lathyrus odoratus* (sweet pea, Figs 56–8), *Cytisus* (broom) and *Wisteria* are common ornamentals. Widespread and common native genera include *Pultenaea* (bush-pea, Figs 60, 61), *Dillwynia* (parrot-pea, Fig. 55; Pl. 7e), *Hardenbergia* (coral-pea) and *Kennedia*.

Most members of the order possess root nodules containing nitrogen-fixing bacteria, and these convert atmospheric nitrogen into nitrogenous compounds that the plant can utilise. Subsequently, any leguminous plant material that is dug into the soil will increase its fertility.

Members of the family are illustrated in Figures 55–61 and Plate 7d–e. For further reading see reference numbers 54, 69, 238, 101 and 236 in the bibliography.

FABACEAE

FLORAL STRUCTURE

Flowers	Zygomorphic, pea-shaped, and sometimes described as papilionaceous (Figs 55–61).
Calyx	Sepals 5, usually united, sometimes with only 3 or 4 obvious lobes, sometimes appearing 2-lipped.
Corolla	Petals 5, 3 of which are free and 2 united. The posterior petal, which is usually the largest, is called the **standard**, the 2 lateral petals are called the **wings** and the 2 anterior petals are more or less united to form the **keel**, a boat-shaped structure that encloses the stamens and carpel (Fig. 58).
Androecium	Stamens 10. Either all are free (Fig. 61), or the filaments are united to form a tube that is open on one side (Fig. 59), or 9 are united with 1 free (Fig. 58).
Gynoecium	Carpel 1. Ovary superior, with 1 to many ovules attached along the posterior side. Placentation marginal.
Fruit	Almost always a legume or pod that is dry at maturity and splits along both sutures to release the seeds. When the pod is constricted between the seeds it is called a lomentum. In *Medicago* (medic, lucerne) the pod is often coiled and when spiny may be referred to as a burr.

Most members of the family are herbs or shrubs, a few are trees, many are climbers. The leaves are simple or compound, and often pinnate or trifoliolate (e.g. clovers), and frequently stipulate. The terminal leaflet or leaflets may be modified to form slender tendrils which will twine around supporting structures (Fig. 56).

SPOTTING CHARACTERS

The 'pea-shaped' or papilionaceous flower. Fruit a legume. Leaves often compound. Stipules often present.

Plant Families

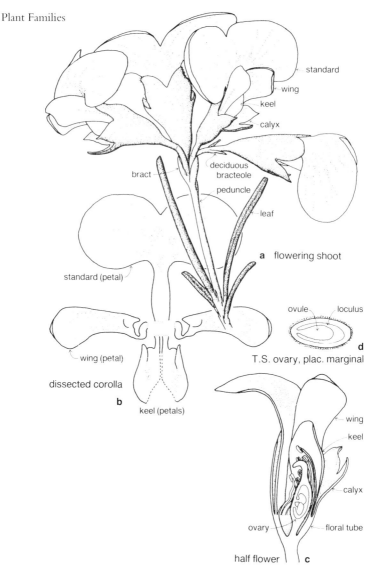

Fig. 55 *Dillwynia glaberrima* (smooth parrot-pea) Fabaceae K(5) C(2),3 A10 G<u>1</u>
Shrub to about 1m tall, stems tough and wiry. Leaves small, terete, about 2 cm long, with very short petioles. Stipules, if present, minute and usually ignored. Flowers yellow, base of petals crimson, in terminal clusters on short lateral branches. Small bracteoles on pedicels deciduous. Small floral tube present. Legume 0.5 cm long. Widespread in heathlands and open forests of Vic., Tas., NSW and SA. Flowering in spring. Spotting characters of the genus: leaves terete, without stipules, bracteoles deciduous, standard broader than long. See also Pl. 7e. (a–b ×4, c ×5, d ×16)

FABACEAE

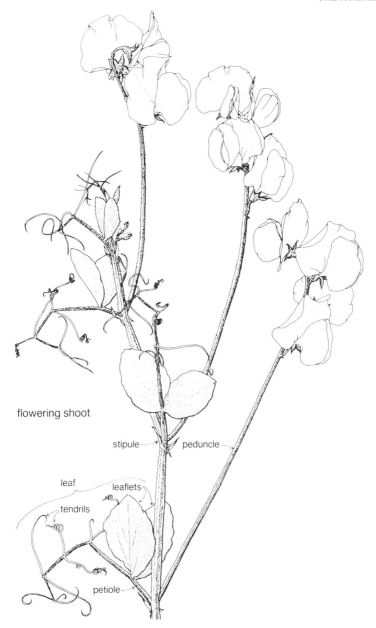

Fig. 56 *Lathyrus odoratus* (sweet pea) Fabaceae
Shoot with compound leaves, and flowers in racemes (×0.6)

Plant Families

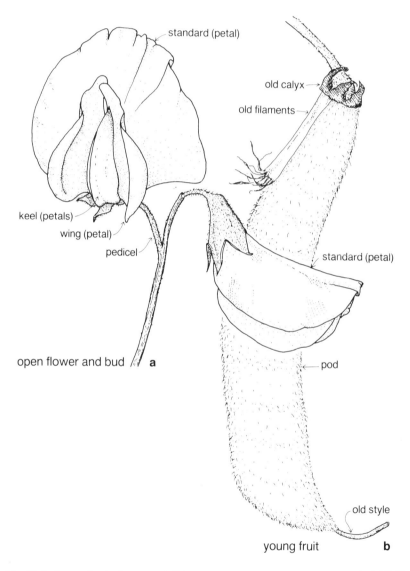

Fig. 57 *Lathyrus odoratus* (sweet pea) Fabaceae (a–b ×1.5)

FABACEAE

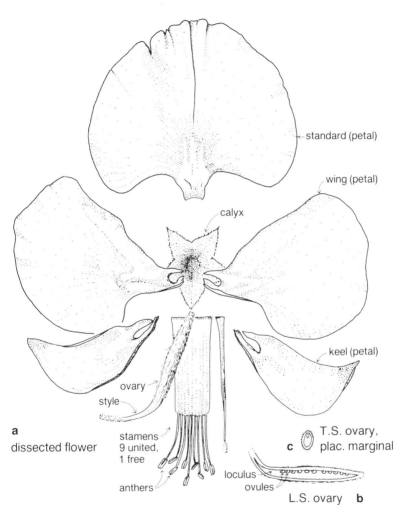

Fig. 58 *Lathyrus odoratus* (sweet pea) Fabaceae   K(5) C(2),3 A(9),1 G1
Annual climber. Leaves compound, with one pair of leaflets up to 5 cm long, the others modified to form tendrils, stipules present. Flowers variously coloured, 1–5 on stout peduncles that are much longer than the leaves. Legume to 6 cm long. Widely cultivated, origin Italy. Flowering in spring. (a–c ×1.5)

Plant Families

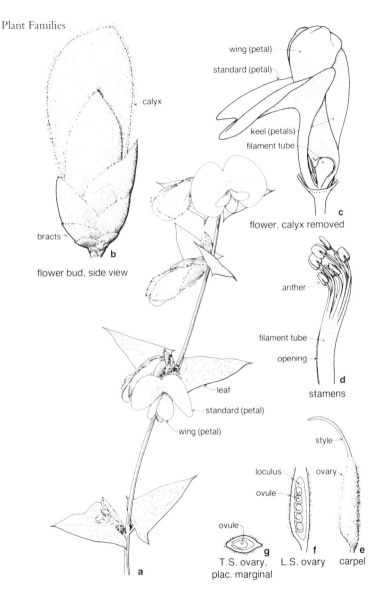

Fig. 59 *Platylobium obtusangulum* (common flat-pea) Fabaceae K(5) C(2),3 A(10) G1
Straggly shrub to about 50 cm. Leaves opposite, ± triangular, about 2–3 cm long, often pungent, stipulate. Flowers dark yellow and red, with brown bracts at the base, 1–3 in upper axils. The 2 lateral calyx lobes are much larger than the other 3, and enclose the flower in bud. Staminal filaments united, but the tube open along 1 side. Legume 1.5–2 cm long. Common in heathlands and drier forests of Vic., Tas., and SA. Flowering in spring. (a ×1, b–f ×4, g ×5)

FABACEAE

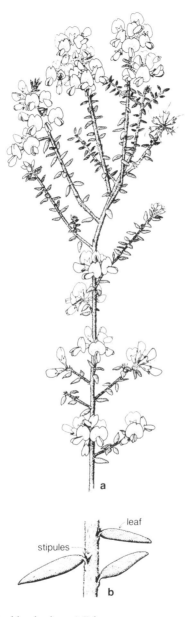

Fig. 60 *Pultenaea gunnii* (golden bush-pea) Fabaceae (a ×0.7, b ×3)

Plant Families

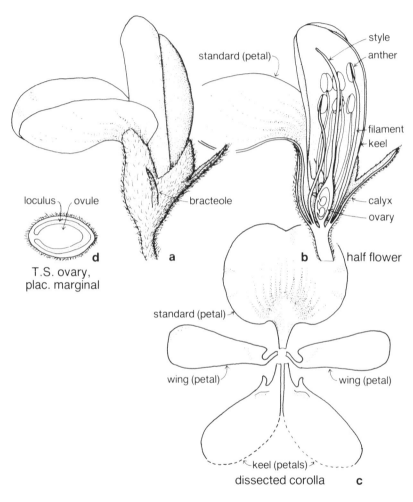

Fig. 61 *Pultenaea gunnii* (golden bush-pea) Fabaceae K(5) C(2),3 A10 G1
Shrub to 1m, profusely branched. Leaves small, ovate, less than 0.5 cm long, stipulate. Flowers orange-yellow, in dense terminal clusters. Bracteoles present on the calyx. Legume 0.5 cm long. Heathlands and drier forests of Vic. and Tas. Flowering in spring. Spotting characters of the genus: stipules, bracteoles on the calyx, standard as broad as long. (a–b ×7, c ×5, d ×20)

RUTACEAE rue family, citrus, boronias, correas

The name Rutaceae is derived from the genus *Ruta*, which includes *R. graveolens* (rue), a European herb supposedly possessed of many healing properties. The family is widely distributed in temperate and tropical regions, especially in Australia and South Africa. The Australian representatives total about 41 genera and 350 species.

Many species are scented, due to the presence of oil glands in their leaves and flowers, and some boronias are grown commercially to provide oils for use in perfumes. Several genera have edible fruits, the most important being the citrus group which includes oranges, lemons, grapefruit and limes. Some Australian natives, *Citrus australasica* (finger lime, formerly *Microcitrus*) and *C. glauca* (formerly *Eremocitrus*, desert lime), bear fruits that are similar to the cultivated citrus and these were used by Aboriginal people and early settlers.

Numerous native and introduced genera are grown as ornamentals—for example *Boronia*, *Correa*, *Philotheca* (previously *Eriostemon*), *Choisya* (Mexican orange) and *Coleonema* (diosma).

Members of the family are illustrated in Figures 62–8 and Plate 1c. For further reading see references 6 and 7 in the bibliography.

FLORAL STRUCTURE

Flowers	Usually actinomorphic and bisexual.
Calyx	Sepals 4–5, free or united, inconspicuous in *Asterolasia*.
Corolla	Petals 4–5, usually free, sometimes united, as in some species of *Correa* (Fig. 63).
Androecium	Stamens usually 8 or 10, in 2 whorls of 4, or 2 whorls of 5. In *Citrus* there are 15 or more. In *Crowea* each anther is tipped with a prominent bearded appendage (Fig. 64).
Gynoecium	Carpels usually 4–5, sometimes free at the base and then united at the top to form a single style and stigma as in *Boronia* (Fig. 62; Pl. 1c) and *Philotheca* (Fig. 65). Ovary superior. Placentation is axile with 1–2 ovules per loculus. A disc is often present at the base of the ovary (Figs 62, 65; Pl. 1c).
Fruit	Commonly dry and leathery, splitting into segments at maturity, or a berry as in the citrus group.

The majority are shrubs or trees and most have alternate leaves that have no stipules. *Boronia*, *Correa* and *Zieria* (Fig. 67) have opposite leaves, which is a useful spotting character for these genera.

Plant Families

SPOTTING CHARACTERS

Plants woody. Leaves contain oil glands and are often pleasantly aromatic when crushed. Other surface features include stellate hairs and peltate scales (Fig. 66). Flowers actinomorphic, 4- or 5-partite, stamens often double the number of petals. Ovary superior, disc often present.

ROUTES THROUGH KEYS

Routes through several keys are given here for members of the family Rutaceae illustrated in this book. Texts are cited by author's name followed by the bibliography number. Note that *Philotheca* (Fig. 65) will key to the older name *Eriostemon* in all but very recent books. In many keys to families, Rutaceae is keyed out more than once as the various genera follow different routes.

BEADLE *et al.* (18)
Key to families, p. 99.
A- dicots -*A-*B-*C-*D-E-F-G-*H-*J-K- group 5 -*A-B- Rutaceae
A- dicots -*A-*B-*C-*D-E-F-G-*H-*J-*K- group 6 -*A-*B-C- Rutaceae

Key to genera, p. 375.
A-B-C-D-E- *Boronia* -A- group 1 -*A-B- *B. mollis*
A-B-C-D-*E- *Zieria* -*A-*C-*E- *Z. arborescens*
A-B-C-*D- *Correa* -A-*B- *C. reflexa*
A-*B-F-G-H-I- *Phebalium* -*A-*D-E- *P. squamulosum*
A-*B-F-G-H-*I-*J- *Eriostemon* -*A-B-C- *E. myoporoides*

BLACK (24)
Key to families, p. 23 (see note about Black's key to families, in the section on Proteaceae). Alternative leads in a couplet are here designated A1, A2, A3; B1, B2 etc.
A2-B2-C2- subclass 1(p. 24) -A3 (p. 26)
 -F2-H4-M1-N1-O1-P2-R1-S1- Rutaceae

Key to genera, p. 492.
A1-B1- *Zieria*
A1-B1- *Boronia*
A1-B2- *Correa* -A2-B1- *C. reflexa*
A2-C1-D2-E1- *Eriostemon*
A2-C1-D2-E2- *Phebalium*

RUTACEAE

WALSH AND ENTWISLE (219)
Key to families, vol. 3, p. 5.
1-2-3-4-9-11-12-13-14-15-16
 -Group 2 -1-2-14-16-17-18-19-20-21-22-23- Rutaceae
1-2-3-4-9-11-12-13-14-15-16
 -Group 2 -1-2-14-16-17-18-19-26-27-28-29-30-32- Rutaceae
1-2-3-4-9-11-12-13-14-15-16
 -Group 2 -1-2-14-16-17-18-19-26-27-28-33-34-35-37- Rutaceae
1-2-3-4-9-11-12-13-14-17- Group 4 -1-2-3-4- Rutaceae

Key to genera, vol. 4, p. 154.
1-2- *Zieria*
1-2-3- *Correa*
1-2-3-4- *Boronia*
1-5-6-8-9- *Eriostemon*
1-5-6-8-10- *Phebalium*

WILLIS (233)
Key to families, vol. 2, p. 1.
1-4-6-7-8-11-88(D)-89-91- Rutaceae
1-4-6-7-8-9-10-23(B)-24-32-34-35-36-37-38- Rutaceae
1-4-6-7-8-9-10-23(B)-24-32-34-35-39-40-41-42-43- Rutaceae
1-4-6-7-8-9-10-23(B)-24-32-34-35-39-40-41-44-45-46-47- Rutaceae

Key to genera, p. 326 (reproduced in Ch. 7, as Key 1).
1-2-3-4- *Correa* -1-3-4- *C. reflexa*
1-2-3-4-5- *Boronia*
1-2-3-4-5- *Zieria* -1-2-3-4- *Z. arborescens*
1-2-3-6-7-8- *Phebalium* -1-6-8-9-11-12- *P. squamulosum*
1-2-3-6-7-8-9- *Eriostemon* -1-2-3-4- *E. myoporoides*

131

Plant Families

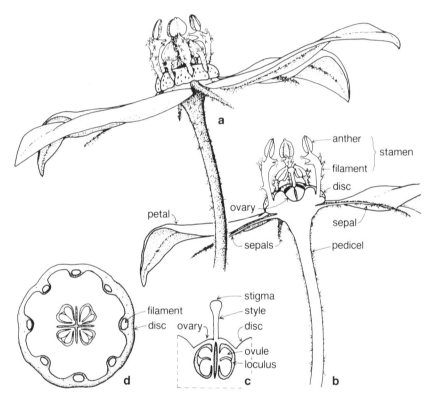

Fig. 62 *Boronia mollis* (soft boronia) Rutaceae K4 C4 A4+4 G($\underline{4}$)
a flower, side view; **b** half flower (the plane of section has passed between the loculi); **c** L.S. carpels (loculi and ovules sectioned); **d** T.S. ovaries, disc and bases of filaments, placentation axile (a–b ×7, c–d ×12)

Small shrub to 2 m tall; branches densely hairy. Leaves opposite, pinnate with 3–7 leaflets. Flowers pink, in small axillary or terminal inflorescences. Stellate hairs on pedicel, calyx and filaments. Outer whorl of stamens longer than the inner. A prominent disc surrounds the ovary and filament bases. Carpels united by the single style, ovaries free. Sandstone gullies inland and south of Sydney, NSW. Often cultivated. Flowering in spring. See also Pl. 1c.

RUTACEAE

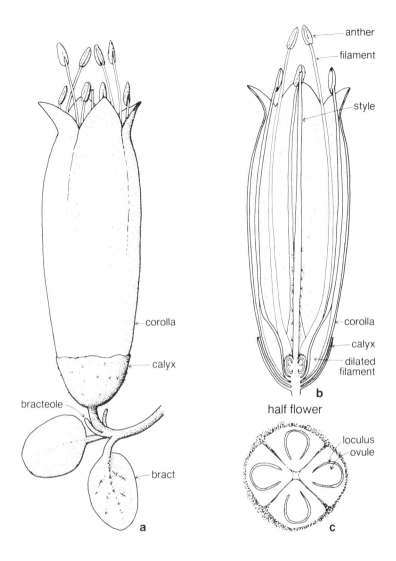

Fig. 63 *Correa reflexa* (common correa, native fuchsia) Rutaceae K(4) C(4) A4+4 G(4)
A variable, small shrub to 2 m tall. Young branches, leaves and flowers with stellate hairs. Leaves opposite, 2–5 cm long, lanceolate to ovate. Flowers usually pendulous, one or a few at the ends of small lateral branches. Colour variable, red, red with green tips, or all green. Bracteoles on pedicel, sometimes falling early. Inner whorl of stamens with broad bases to filaments. Carpels united, ovaries separating at maturity. Occurs in all states, common. Flowering late autumn to spring. (a–b ×2.5, c ×13)

Plant Families

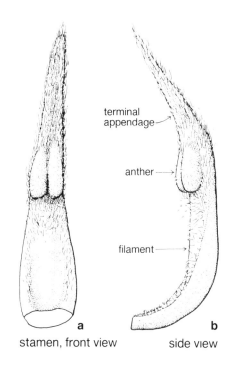

Fig. 64 *Crowea* (crowea) Rutaceae
Crowea is a small Australian genus of pink-flowered shrubs, including three species and several cultivars that are commonly grown. It is closely related to *Philotheca* and has the same floral formula. One of the key differences is the hairy anthers: note the prominent, bearded, terminal appendage. (a–b ×12)

RUTACEAE

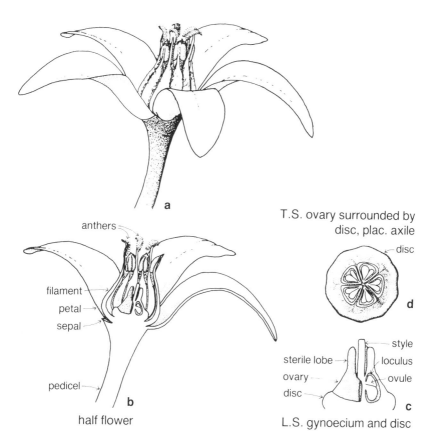

Fig. 65 *Philotheca myoporoides* (long-leaf wax-flower) Rutaceae   K5 C5 A5+5 G(5)
Variable shrub to 5 m tall. Stems glabrous, rough due to prominent oil glands. Leaves alternate, lanceolate, aromatic. Buds pink, flowers white in stalked axillary umbels. Filaments hairy but anthers glabrous, tipped by a small point. Ovaries united by the base of the single style, each ovary with a prominent terminal lobe. Disc prominent around the base of ovaries. Vic., NSW, Qld. Often cultivated, flowering spring and summer. Formerly known as *Eriostemon myoporoides*. (a–b ×5, c–d ×10)

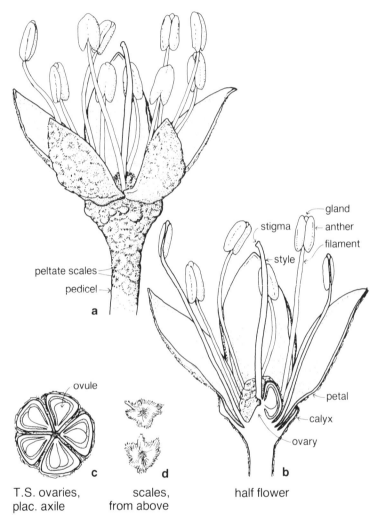

Fig. 66 *Phebalium squamulosum* (forest phebalium) Rutaceae   K(5) C5 A5+5 G(5)
Erect shrub of variable form. Leaves alternate, narrow oblong to elliptical, 1–7 cm long. Most parts of the plant covered with silvery or brownish, overlapping, peltate scales. Flowers cream to yellow, borne in small terminal umbels. Each anther bears a small, rounded, apical gland. Ovaries densely scaly, united at the base and by the single style. Forests of eastern Vic., NSW and Qld. Cultivated, flowering in spring. (a–b ×10, c–d ×20)

RUTACEAE

Fig. 67 *Zieria arborescens* (stinkwood) Rutaceae (×0.7)

Plant Families

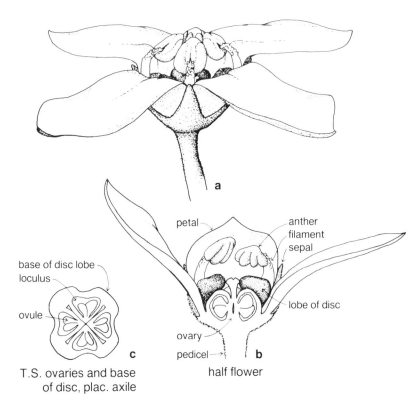

Fig. 68 *Zieria arborescens* (stinkwood) Rutaceae   K4 C4 A4 G($\underline{4}$)
Tall shrub or small tree to 5 m. Leaves opposite, trifoliolate, leaflets 3–10 cm long with an unpleasant odour. Branchlets and undersides of leaves covered with minute stellate hairs. Flowers white, small, borne in axillary cymose panicles. Disc 4-lobed. Stamens with warty filaments. Damp forests and gullies of Tas., Vic. and NSW. Flowering late winter to early summer. Occasionally cultivated. (a–c ×10)

EUPHORBIACEAE spurges

The Euphorbiaceae is a very large family of trees, shrubs and herbs, and is predominantly tropical. A number are of economic importance, including *Hevea brasiliensis*, the source of natural rubber, *Manihot esculenta*, the cassava or tapioca plant, and *Ricinus communis*, from which castor oil is extracted.

The large genus *Euphorbia*, of about 2000 species, includes herbs and shrubs of very diverse appearance, but all produce a milky juice that is often poisonous. The inflorescence, sometimes called a cyathium, consists of several male flowers and 1 female flower enclosed in a cup-shaped involucre of 5 fused bracts. In the small herbaceous annual *E. peplus* (petty spurge, Fig. 69) there is a crescent-shaped gland at 4 of the 5 junctions between the bracts. Within the involucre, the female flower consists of a trilocular superior ovary, and the male flowers of 1 stamen only. There are no perianth parts. The stamens have a jointed stalk, and the part below the joint is interpreted as a pedicel, with the filament above. The female flower hangs over the part of the involucre that has no gland. Each loculus contains 1 ovule and there are 3 forked styles. A number of euphorbias are grown as ornamentals, including *E. wulfenii* (spurge, Pl. 6a) and *E. pulcherrima* (poinsettia), which has bright red floral leaves surrounding the inflorescences. *E. peplus* is a widespread weed native to Europe.

Ricinocarpos pinifolius (wedding bush, Fig. 70) is a bushy shrub to 1.5 m tall, with opposite, narrow-linear leaves 1–3.5 cm long, that have revolute margins. It is found in the coastal heaths of Vic., Tas. and NSW, and differs from most members of the family in having conspicuous petals. The flowers are actinomorphic and unisexual, and the females are either solitary or surrounded by a cluster of males. The calyx is of 4–6 united sepals and there are 4–6 free petals. In the male flowers the numerous stamens are united by their filaments forming a central column. In the females the ovary is superior with 3 loculi, and it is covered by short stout hairs. Each of the 3 styles is deeply divided.

Another distinctive member of the family is *Amperea xiphoclada* (broom spurge), an almost leafless, small, tufted shrub with triangular stems, found in heathlands and poor forest country.

Plant Families

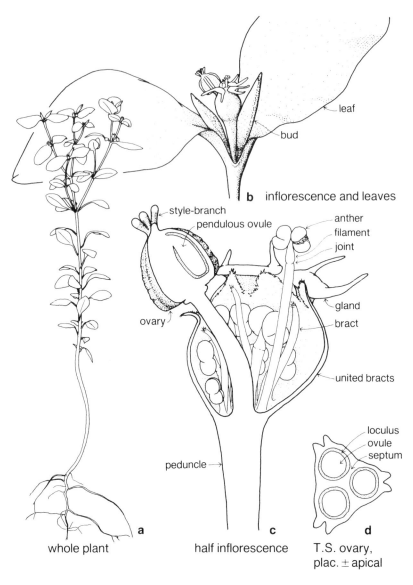

Fig. 69 *Euphorbia peplus* (petty spurge) Euphorbiaceae
(a ×0.7, b ×7, c–d ×30)

P0 A0 G(<u>3</u>) ♀
P0 A1 G0 ♂

140

EUPHORBIACEAE

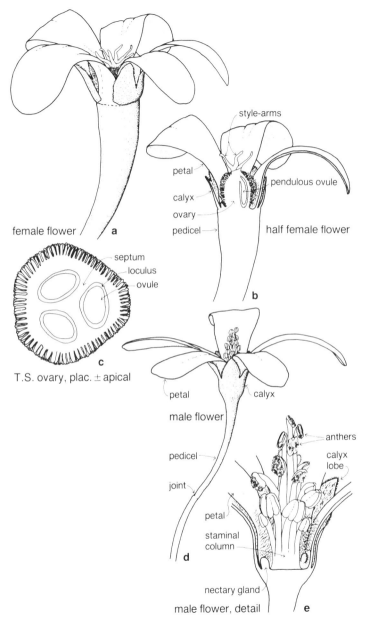

Fig. 70 *Ricinocarpos pinifolius* (wedding bush) Euphorbiaceae (a–b ×3, c ×12, d ×3, e ×8)

K(5) C5 A0 G(3) ♀
K(5) C5 A(∞) G0 ♂

STACKHOUSIACEAE

This is a small family of three genera, almost entirely restricted to Australia.

The largest genus is *Stackhousia* with about 20 species, found in all states. These small annual or perennial herbs usually have small alternate leaves. The flowers are actinomorphic, bisexual and perigynous, and the inflorescence is often a spike or raceme. The short floral tube (Fig. 71) bears 5 sepals and has a tubular corolla of 5 white to yellowish petals, which are usually united in the middle but free both at the base and the top. The 5 stamens are of unequal length. There are 2–5 united carpels, often 3, with the same number of loculi in the superior ovary and 1 ovule per loculus. At maturity the fruit breaks into one-seeded nutlets. The flowers are particularly fragrant at night, and some are pollinated by moths.

STACKHOUSIACEAE

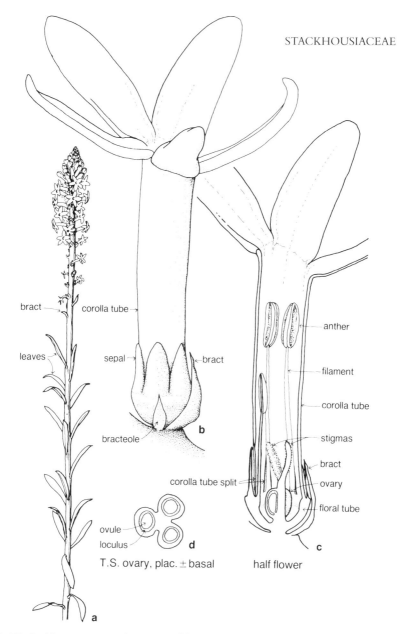

Fig. 71 *Stackhousia monogyna* (creamy stackhousia) Stackhousiaceae
K5 C(5) A5 G(3) floral tube present
Small, erect, perennial herb about 40 cm tall, with a few branches from near the base. Widespread in south-eastern Australia, flowering in spring and early summer. (a ×0.6, b–c ×12, d ×24)

RHAMNACEAE buckthorn family

The family is a large one, with worldwide distribution, and 25 of its 60 genera are Australian. The family name is derived from the genus *Rhamnus*, and *R. catharticus* (buckthorn) is used extensively as a hedge plant in Europe. *R. alaternus* (Italian buckthorn) is naturalised in south-eastern Australia.

Pomaderris (Fig. 73) and *Spyridium* (dusty miller) are common genera in south-eastern Australia, and both are grown in native gardens.

FLORAL STRUCTURE

Flowers   Usually actinomorphic and bisexual. Floral tube often present.
Calyx   Sepals 4–5, free. The floral tube often resembles a tubular calyx.
Corolla   Petals 4–5, sometimes hooded over the stamens as in *Spyridium* and *Cryptandra* (Fig. 72), often absent, e.g., in some species of *Pomaderris* (Fig. 73).
Androecium   Stamens 4–5, alternating with the sepals.
Gynoecium   Carpels 1–5, united. Ovary usually inferior, sometimes semi-inferior or superior, depending on the degree of union of the ovary wall and floral tube. A nectary disc often lines the inside of the floral tube and/or part of the ovary. Ovule 1 per loculus.
Fruit   Usually dry, splitting into 1-seeded sections. Sometimes a drupe.

The majority of family members are small trees or shrubs. Leaves are simple, often rugose (wrinkled) or with a rough surface. Stipules are present, but may be deciduous. Stellate (star-shaped) hairs are often found on leaves and young stems.

SPOTTING CHARACTERS

Plants woody, young stems and leaves often with stellate hairs. Leaves rough or rugose, stipulate. Petals (if present) opposite the stamens and often hooded over them. Ovary often semi-inferior.

RHAMNACEAE

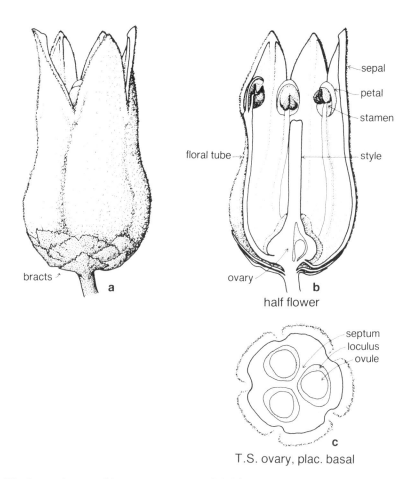

Fig. 72 *Cryptandra amara* (bitter or spiny cryptandra) Rhamnaceae

K5 C5 A5 G(<u>3</u>) floral tube present

Shrub, less than 1 m high, with rigid branches, branchlets covered with stellate hairs and often ending in a spine. Leaves small, ± oblanceolate, 2–6 mm long, flat or with recurved margins. Flowers white, solitary or in clusters of 2–3. Petals small, hooded over the stamens. Fruit a capsule separating into 3 fruitlets. Sepals and floral tube persistent in fruit. Widespread in dry forests and heathlands of Vic., SA, Tas., NSW and Qld. Flowering in winter and spring. (a–b ×12, c ×25)

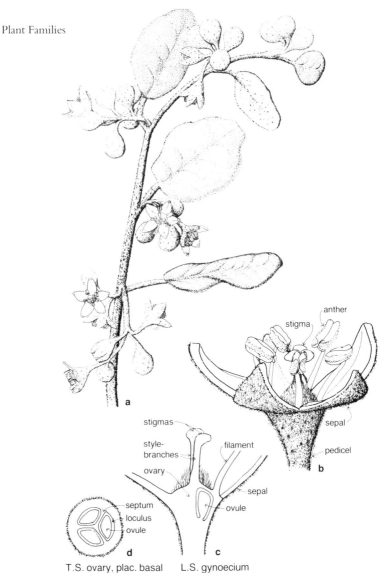

Fig. 73 *Pomaderris paniculosa* ssp. *paralia* (coast pomaderris) Rhamnaceae

K5 C0 A5 G(-3-)

Shrub 1–2.5 m high. Young stems and leaves covered with brown stellate hairs. Leaves ovate, 1.5–5 cm long, dark green and glabrous on top with a whitish margin, white hairs underneath and brown stellate hairs along the veins. Flowers yellowish-white to pink, in axillary racemes or small panicles. Stamens alternate with sepals. Style deeply divided. Fruit a capsule separating into 3 fruitlets. Widespread along the coastline of all southern states, particularly on limestone cliffs. Flowering in spring. (a ×2, b–d ×10)

STERCULIACEAE kurrajongs, paper-flowers

A large, mainly tropical family of trees, shrubs and herbs. The type genus *Sterculia* is native to northern Australia. The closely related *Brachychiton* (kurrajong) is found in all states except Tasmania. *Thomasia* (paper-flower) and its close relative *Lasiopetalum* (velvet-bush) are widespread Australian genera.

Thomasia petalocalyx (Fig. 74) is a low coastal shrub found in western Vic., SA and WA. The flowers are bisexual and usually actinomorphic, with 1–3 sepal-like bracteoles subtending the petaloid calyx. The 5 sepals are united at the base, and mauve–lilac in colour, with a conspicuous midrib. Petals are often absent but if present are small, dark red and gland-like, and opposite the stamens. The 5 stamens have dark red anthers that dehisce through a terminal pore or short slit. The ovary is superior, with 3 loculi, each containing 2 or more ovules.

The stems, undersides of leaves and bracteoles are covered with brownish stellate hairs. The leaves are ovate to oblong, truncate and stipulate, and often have undulate, slightly recurved margins and conspicuous veins. The stipules are large and rounded, and may be mistaken for small leaves. *Lasiopetalum* differs in having no stipules and free sepals.

ROUTES THROUGH KEYS

Routes through several keys are given here for *Thomasia petalocalyx* (Fig. 74) to family, genus or species level depending on the extent of the key. Texts are cited by author's name followed by the bibliography number.

BLACK (24)

Key to families, p. 23 (see note about Black's key to families, in the section on Proteaceae). Alternative leads in a couplet are here designated A1, A2, A3; B1, B2, etc.
A2-B2-C2- subclass 1(p. 24) -A3(p. 26)-F2-H6(p. 28)-J2- Sterculiaceae

Key to genera, p. 569.
A1-B3- *Thomasia* - *T. petalocalyx*

GEORGE/ORCHARD (94)

Key to families, vol. 1, 1st edn, p. 113.
1-2-3-10-572-647-652-653-667-719-720-721-722-723-724-725-726-727
 -728- Sterculiaceae

Key to families, vol. 1, 2nd edn, p. 521.
1-2-3-10-693-777-782-783-802-812-813-869-870-871-872-873-874-875
 -877-878-879- Sterculiaceae

Plant Families

JESSOP AND TOELKEN (127)

Key to families, vol. 1, p. 65. The second lead of each pair is here designated by ' following the number.
1'-3'-4- Dicotyledonae (to p. 66)
   1'-2'-3'-8'-9-10'-32'-33'-43'-47'-53'-61'-62'- Sterculiaceae.

Key to genera, vol. 2, p. 848.
1'-2-3'- *Thomasia*- *T. petalocalyx*

WILLIS (233)

Key to families, vol. 2, p. 1.
1-4-6-7-12-138(F)-144-145-146-147-148-149-150-151- Sterculiaceae

Key to genera, p. 382.
1-2-4- *Thomasia* - *T. petalocalyx*

STERCULIACEAE

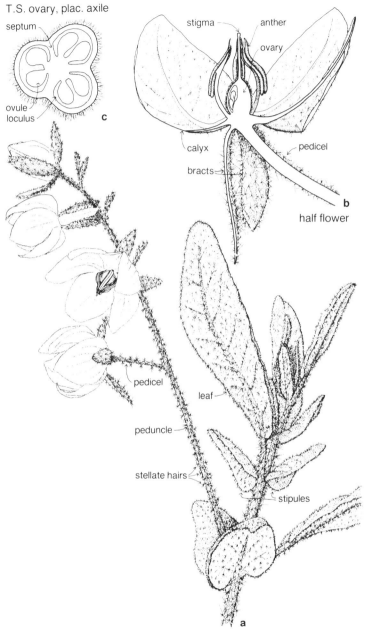

Fig. 74 *Thomasia petalocalyx* (paper-flower) Sterculiaceae K(5) C0 A5
G(3)

DILLENIACEAE guinea-flowers

This family of trees, shrubs and climbers is almost entirely tropical. The type genus *Dillenia* is found in northern Australia and extends to South-east Asia.

In southern Australia the genus *Hibbertia* (guinea-flower) is widespread and the bright yellow flowers that are characteristic of the family make them conspicuous components of heathland and forests.

The flowers of *Hibbertia* (Fig. 75; Pl. 1b) are usually regular and bisexual. There are 5 persistent sepals. The 5 petals are often crumpled in bud, and may fall soon after anthesis. The few to numerous stamens are free or partly united at the base, and some may be reduced to staminodes. They are either in a single group on one side of the carpels or in a ring around them. The 2–3 carpels are free, often with curved styles, and contain 1–many ovules. The fruit is a cluster of follicles.

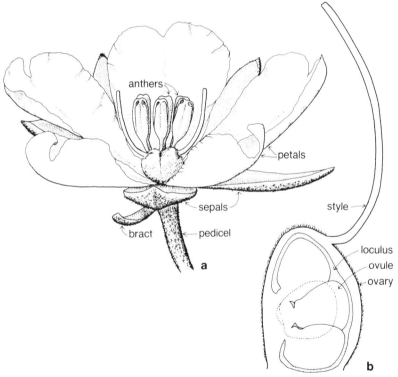

Fig. 75 *Hibbertia riparia* (guinea-flower) Dilleniaceae   K5 C5 A6 G$\underline{2}$

Erect shrub to about 50 cm tall. Leaves linear-oblong, to 8 mm long, often with stellate hairs, margins revolute, apex obtuse. Inner surfaces of sepals shiny, outer surfaces with stellate hairs. This variable species is widespread in southern and eastern Australia; this variant occurs in southern Vic. and is also recorded in SA. Flowering in spring. This variant is listed in Willis (bibliography no. 233) as *H. australis*. See also Pl. 1b. (a ×7, b ×25)

THYMELAEACEAE rice-flowers

The Thymelaeaceae is a small family of trees and shrubs found in the temperate and tropical regions of the world. The genus *Thymelaea*, after which the family is named, is native to Asia and the Mediterranean, and one species has been introduced to South Australia and Victoria. *Daphne*, with its fragrant flowers, is a common garden shrub.

In Australia, an interesting member of the family is *Drapetes tasmanica*, one of the so-called cushion plants of the Tasmanian high country. The largest genus, *Pimelea* (rice-flower), is widespread, with some 90 species being found in a range of habitats.

The flowers of *Pimelea* (Fig. 76) are small, bisexual or unisexual, and commonly clustered in terminal heads surrounded by 4 or more involucral bracts. The 4 white or yellow sepals extend from the top of the slender floral tube. Petals are absent, and the 2 stamens are inserted near the top of the tube. Sometimes this arrangement is interpreted as a calyx tube with episepalous stamens, and in other genera extra small 'scales' are variously described as petals or staminodes. The ovary is superior, with a single loculus containing 1 pendulous ovule. The fruit is a drupe or nut that is enclosed in the persistent lower part of the floral tube.

The leaves of *Pimelea* are entire, often small, opposite or alternate and without stipules. The bark on the stems and branches of some species such as *P. axiflora* (bootlace bush) is very strong, and early settlers often used strips of it in place of string or twine.

For descriptions and keys to all Australian species, see *Flora of Australia* vol. 18 (bibliography no. 94).

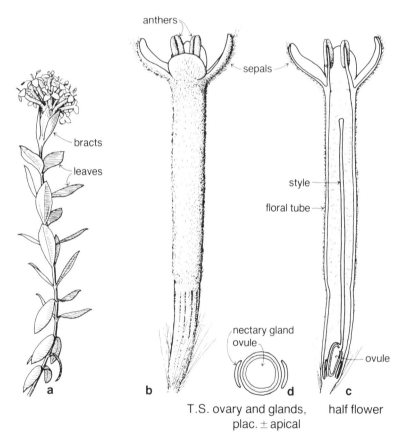

Fig. 76 *Pimelea glauca* (smooth rice-flower) Thymelaeaceae

K4 C0 A2 G$\underline{1}$ floral tube present

Small bushy shrub, ± 0.5 m tall, with opposite leaves. Flowers creamy white, sometimes unisexual, in terminal heads surrounded by 4 bracts (the inner 2 with ciliate margins). Widespread in SA, Vic., Tas., NSW and Qld. Flowering in spring. (a ×1, b–c ×7, d ×20)

MYRTACEAE myrtle family, eucalypts, bottlebrushes, tea-trees

This large family is mainly of the southern hemisphere and the tropics, with an extension to South-east Asia. About half of the 155 genera are native in Australia, 45 in Central and South America, but only four in Africa. The genus *Myrtus* (myrtle), from which the family name is derived, is found in the Mediterranean region.

The family was very important to Aboriginal people. They collected water from the roots of many mallee eucalypts, edible grubs and insects from the foliage or under the bark of many species, and used the wood and bark of eucalypts and melaleucas for a variety of purposes.

The genus *Eucalyptus* has considerable economic importance, as many species are utilised for timber, chipboard, paper-making or extraction of aromatic oils, and they are also important to bee-keepers. *Acmena* (previously *Eugenia*, lillypilly), *Lophostemon* (brush box) and *Angophora* (apple box) are eastern Australian trees, sometimes used to provide timber for cabinet work and, like *Eucalyptus* are widely planted as street trees and ornamentals. *Feijowa sellowiana* (pineapple guava) and *Psidium guajava* (guava), introduced from South and Central America respectively, produce delicious fleshy fruits. The dried unripe berries of *Pimenta dioica* of the West Indies are sold as allspice and the dried flower buds of *Syzygium aromaticum* from the Moluccas are marketed as cloves.

Many well-known ornamentals include *Agonis* (willow myrtle), *Callistemon* (bottlebrush), *Chamelaucium* (wax flower), *Leptospermum* (tea-tree), *Melaleuca* (paper-bark or honey-myrtle) and *Thryptomene*. Many countries in South America, Africa and the Mediterranean region, and also California and New Zealand, have successfully introduced eucalypts for use in plantations and as street trees.

Myrtaceae is divided into two subfamilies; the predominantly Australian Leptospermoideae, most of which have dry capsular fruits, and the mainly South American Myrtoideae, with succulent fruits. Most southern Australian members of the family belong to the Leptospermoideae, one of the exceptions being *Acmena* (lilly-pilly). Members of the family are illustrated in Figures 77–87 and Plates 3d and 6f.

For further reading on various genera see bibliography numbers 26, 114 and 240.

Plant Families

FLORAL STRUCTURE

Flowers	Nearly always actinomorphic and bisexual.
Calyx	Sepals 4–5, usually free, united in *Eucalyptus*.
Corolla	Petals 4–5, usually free, united in *Eucalyptus*.
Androecium	Stamens usually numerous, occasionally 5 or 10, sometimes united in bundles, for example *Melaleuca* (Fig. 87; Pl. 6f) and *Lophostemon*.
Gynoecium	Carpels usually 2–10, commonly 5, united. Usually the number of loculi equals the number of carpels, but occasionally there is only 1 loculus as in *Calytrix* (Fig. 79), *Thryptomene*, and several other genera. Ovary usually inferior, sometimes semi-inferior or superior. A floral tube, the length of which varies with species, is fused with and may extend above the ovary and bears the other whorls of parts. The inside of the tube is often lined with nectar-producing tissue. Placentation usually axile. Ovules 1 or 2–many per cell.
Fruit	Often a woody capsule, opening by valves at the top, or a berry.

Differences in the proportions of petals and stamens can disguise the real similarities in structure. For example, in *Leptospermum* (tea-tree, Fig. 86), the petals form the showy part of the flower and the stamens are relatively short. In the bottlebrushes, the long showy stamens are conspicuous and often obscure the petals (*Callistemon*, Fig. 78).

Members of the family are woody shrubs and trees. Leaves are alternate or opposite, simple and with no stipules. The leaves, and often other parts of the plant, are dotted with glands containing aromatic oil.

SPOTTING CHARACTERS

Plants woody. Flowers regular, ovary often inferior. Fruit usually dry, often a woody capsule. Oil glands present in the leaves, usually distinctively aromatic when crushed.

THE GENUS *Eucalyptus*

Eucalyptus is the largest genus in the Myrtaceae, with many hundreds of species, almost all of which are restricted to Australia. They are spread throughout the continent in a wide range of habitats and climatic conditions, and are the chief components of the forests of south-western, south-eastern and eastern Australia. Many species, such as *E. pauciflora* (snow gum), survive freezing temperatures, while others, like the mallees, are adapted to high temperatures and drought.

MYRTACEAE

Classification

The first *Eucalyptus* species, *E. obliqua* (messmate) was described in 1789, from specimens collected on Bruny Island, southern Tasmania, during Cook's third expedition. By 1866 George Bentham, in his *Flora Australiensis*, accepted 135 species, grouping them into five series based on features of the stamens. Later botanists also relied on the stamens (particularly the anthers) in devising systems of classification although the weighting of a particular set of characteristics such as this is now considered unsatisfactory. In practice, anther types can be difficult to observe even in young flowers. The three main anther types are
- macrantherous—anther lobes parallel, dehiscence by parallel slits
- renantherous—anthers reniform (kidney-shaped), dehiscence by slits
- porantherous—anthers dehisce by terminal pores (Fig. 84d).

Anther characters are described, illustrated and used in keys in a recent field guide (bibliography no. 37, vol. 1, first edn).

In the twentieth century, many botanists contributed to the understanding of the genus, which has come to be regarded as a taxonomically difficult group currently including over 700 species.

In 1995 botanists from the New South Wales Herbarium published a proposal to remove the 'bloodwoods' from the genus *Eucalyptus* and place them in a new genus, *Corymbia*. This was part of a formal move to reclassify eucalypts, and followed many years of study, discussions and proposals, some of which, although published, were nevertheless intended only as informal reviews. At various times up to nearly a dozen genera have been suggested as warranting recognition alongside a more restricted interpretation of *Eucalyptus*.

Most bloodwoods are native to northern Australia but examples common in cultivation in the southeast include *Corymbia ficifolia* (syn. *Eucalyptus ficifolia*, Western Australian flowering gum) and *C. citriodora* (syn. *E. citriodora*, lemon scented gum). The natural ranges of *C. gummifera* (bloodwood) and *C. maculata* (spotted gum) extend south into Victoria in the far east.

The genus *Corymbia*, with 113 species, is characterised by the combination of a number of rather technical features including well-developed tertiary venation in the leaves, and the presence of elongated bristle-glands. Several microscopic and anatomical features are also important. These features are shared by the closely related genus *Angophora* from which *Corymbia* is distinguished by its possession of an operculum, and 'alternate' adult leaves. In addition, corymbias bear their flowers in 'terminal panicles' which often results in the canopies being conspicuous at flowering time. They also produce urn-shaped fruits, their leaves have secondary veins at a large angle to the midrib, and the bark is often distinctively rough and flaky (tessellated). However, as happens in many plant groups, not all species possess all of these features (i.e., exceptions occur), and

Plant Families

further, some of these characteristics occur in other eucalypts. This situation (overlap of 'distinctive' characters) occurs occasionally in classification, and it is up to the botanist, having weighed all the evidence, to make the decision as to where to draw the lines between the various groups.

It should be remembered that plant classification is continually refined over time with the benefit of more information and increasing sophistication in techniques. As well, interpretation of the evidence is sometimes not unanimous—opinions differ and botanists may hold different views, particularly over the ranks to be assigned to various groups. With respect to the genus *Eucalyptus*, an example of this situation is the recently published proposal to *broaden* the generic concept (to include the closely related genus *Angophora*) rather than to subdivide it into a number of smaller genera. In this case those species now in *Angophora* would change their names to *Eucalyptus*, and of course those species assigned to *Corymbia* would remain in *Eucalyptus*.

For the non-expert, perhaps the main question now becomes 'Well, what proper name do I apply to the Western Australian flowering gum?' Whether one uses *Corymbia ficifolia* or *Eucalyptus ficifolia* is really a matter of personal choice, realising that behind the use of either of these names is the acknowledgement of one alternative view of the classification of the group. In time, one view of the classification of eucalypts will probably prevail, and the use of one or other of the scientific names for the Western Australian flowering gum will come to be accepted.

The structure of the open flowers of *Eucalyptus* is relatively uniform, and the identification of species is largely based on characters of the buds, fruits, leaves and bark. The following notes refer to *Eucalyptus* as it is traditionally known, i.e. including *Corymbia* but excluding *Angophora*.

Buds

The eucalypt bud has three main parts: a pedicel, which may be absent, floral tube (variously called a calyx tube, thalamus tube or torus) and an operculum (Figs 80, 83). The **operculum**, which is deciduous, is formed by one or two whorls of united perianth parts. Some species with a double operculum shed the outer or sepaline one as the bud develops, leaving a circular scar on the outside of the bud. A scar is not seen when the opercula fall together or when the operculum is single.

Bud characters used in identification include:
- whether the bud is pedicellate or sessile
- length of the floral tube measured from the top of the pedicel, relative to the operculum
- shape of operculum—conical, rounded, beaked etc.

- surface characters—whether smooth, glaucous, warty, ribbed etc.
- inflorescence type—buds may be solitary or arranged in umbels (Figs 80, 83) or panicles; the number of buds per umbel, and characters of the peduncle are also used.

Flowers

The eucalypt flower has numerous stamens arranged in several whorls on the floral tube. In bud, the stamens are either all inflexed towards the style, all erect, or inflexed but with the anthers lying irregularly around the style. At anthesis the filaments straighten out so that all the stamens are more or less erect. In some species the outermost stamens are replaced by staminodes (Fig. 84).

Fruit

The fruit is a woody capsule with usually 3–5 **valves** through which the seeds are released. The number of valves may vary even in fruits from the same tree. At the fruiting stage the opercular scar is known as the **rim**, and the scar left by the stamens, the **staminal ring**, may also be distinct. Sometimes the particular area of tissue where the stamens were attached persists as a withered brown ring on the developing fruit (e.g., *E. leucoxylon*, yellow gum, Fig. 84b); in some texts it is referred to as a staminophore. The tissue between the rim and the base of the valves is the **disc**, and this may be flat, ascending (Fig. 80) or descending (Fig. 83). The valves open outwards and their position in relation to the rim is described as level, exserted (Fig. 80) or enclosed (Fig. 83).

Leaves

The leaves of most eucalypts change as the plant grows from seedling to adult form (Fig. 82). Juvenile leaf characters are sometimes used to distinguish between closely related species such as *E. rubida* (candlebark) and *E. viminalis* (manna gum). The adult leaves of most species are tough and leathery and tend to hang vertically from the branches. Leaves form in opposite pairs but in the majority of species adult leaves finally appear alternate due to unequal elongation of the axis between the leaf bases. A notable exception is *Eucalyptus crenulata* (Buxton gum), common in cultivation, which retains the distinctive juvenile leaf phase on the mature tree. Most of the aromatic oils found in the oil glands of the leaves have a 'eucalyptus' odour, even though the constituent oils vary with species. Many eucalypt leaves are slightly sickle-shaped, others are asymmetrical, with the two halves of the lamina joining the petiole at different points. This feature of its leaves led to the name *E. obliqua* (messmate), a widespread forest tree in southern Australia.

Bark

The type of bark is a considerable help in the identification of species, particularly in the field. The main categories are nonpersistent and persistent. In the non-persistent type, the outer layer of bark is shed each season, leaving the trunk and branches smooth and light-coloured. This type is often referred to as 'gum bark' and may be shed completely, in irregular patches, or, for a time, hang from the branches in long ribbons.

In the persistent type, the outer layer is not shed regularly but tends to flake off gradually, so the surface of the trunk is rough and mostly dark-coloured. Several variations are recognised:
- stringybark—thick, often furrowed, with long coarse fibres; the bark can be pulled off in strips
- peppermint—fibrous but relatively compact; fibres of medium length, not easily pulled off
- box—thin, compact, short-fibred, sometimes breaking horizontally into irregular flakes
- ironbark—deeply furrowed, dark and hard due to gum deposition
- bloodwood—tessellated (that is, cracking vertically and horizontally to form small plates).

Form classes

The majority of eucalypts can be divided into three general form classes:
- Forest trees have a tall single stem with small branches high above the ground forming a relatively small crown.
- Woodland trees have a relatively short, single stem, before branching to form a spreading crown.
- Mallees have many thin stems arising at ground level. The stems are usually less than 10 m high, with a small crown.

Recovery after fire

At or below ground level, most eucalypts develop a swollen woody structure called a **lignotuber**, which contains food reserves and many dormant vegetative buds. These buds rapidly develop into new shoots when the aerial parts of the plant are destroyed, for example, by fire or harvesting for oil distillation. Mallee eucalypts have particularly well-developed lignotubers, which are collected from cleared land and sold for fuel as mallee roots. Eucalypts also recover from fire in other ways. When only the crown and small branches are killed, concealed buds just under the bark of the trunk and branches sprout to form a mass of epicormic shoots. When the crown has been re-established most of these shoots die off. This type of regeneration is typical of the stringybarks, though not restricted to them. Species with thin bark and no lignotuber, such as *E. regnans* (mountain ash), are readily killed by fire and must regenerate from seed. The capsules, which protect the seeds from all but very severe fires, release them soon after the tree has died, and when conditions are suitable a thick growth of seedlings will appear.

There are several useful guides for eucalypts, see bibliography numbers 37, 58, 146 and 160. On the particular issue of eucalypt classification see 35, 86, 111, 179. The *Flora of Australia* vol. 19 (bibliography no. 94) includes numerous line drawings, keys and descriptions for all species recognised at the time of publication (1988), but there has been much research into the genus since then and numerous new species have been described. The *Field Guide to Eucalypts* (no. 37) is probably the most comprehensive and up-to-date text. *Native Trees and Shrubs of South-eastern Australia* (no. 57) is a valuable introduction to the species of the region, and the family in general. See the list at the beginning of the bibliography for other references.

Plant Families

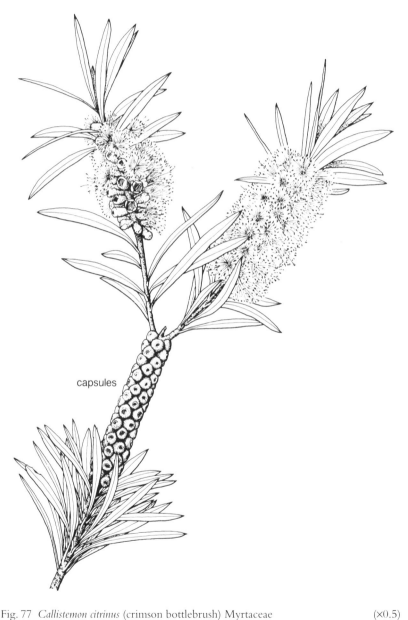

Fig. 77 *Callistemon citrinus* (crimson bottlebrush) Myrtaceae (×0.5)

MYRTACEAE

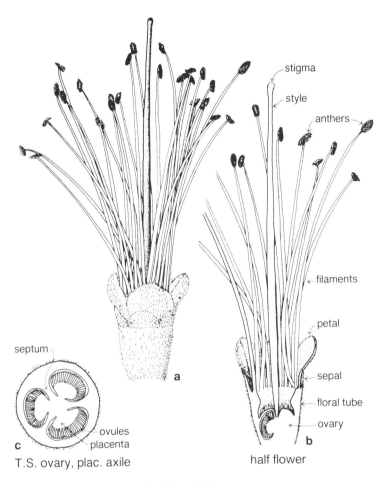

Fig. 78 *Callistemon citrinus* (crimson bottlebrush) Myrtaceae

K5 C5 A∞ G($\overline{3}$) floral tube present

Shrub 1–3 m tall, branches often arching over. Young stems and leaves covered with deciduous, silky hairs. Leaves lanceolate, up to 8 cm long. Flowers red, borne in terminal spikes. The apical bud remains vegetative and the shoot continues to grow beyond the inflorescence. Fruit a capsule, often retained on the plant for many years. Common in swampy heaths of eastern Vic., NSW and Qld. Widely cultivated and many cultivars and hybrids are known. Flowering in spring to early summer, sometimes autumn. (a–b ×4, c ×7)

Plant Families

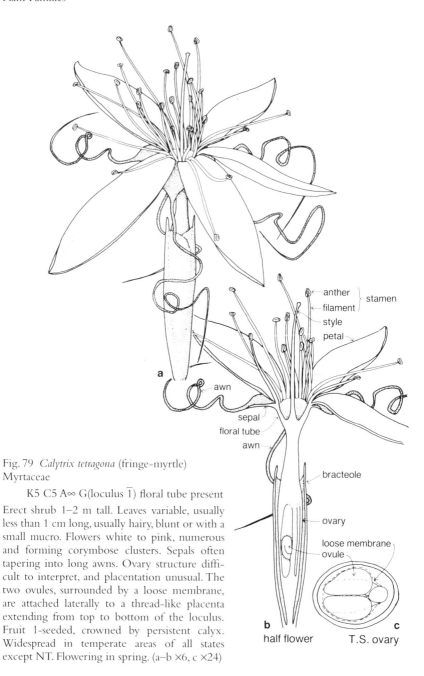

Fig. 79 *Calytrix tetragona* (fringe-myrtle) Myrtaceae

K5 C5 A∞ G(loculus $\bar{1}$) floral tube present

Erect shrub 1–2 m tall. Leaves variable, usually less than 1 cm long, usually hairy, blunt or with a small mucro. Flowers white to pink, numerous and forming corymbose clusters. Sepals often tapering into long awns. Ovary structure difficult to interpret, and placentation unusual. The two ovules, surrounded by a loose membrane, are attached laterally to a thread-like placenta extending from top to bottom of the loculus. Fruit 1-seeded, crowned by persistent calyx. Widespread in temperate areas of all states except NT. Flowering in spring. (a–b ×6, c ×24)

162

MYRTACEAE

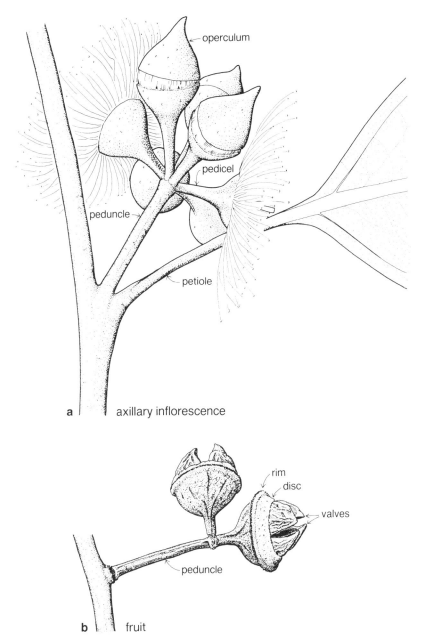

Fig. 80 *Eucalyptus camaldulensis* (river red gum) Myrtaceae (a–b ×4)

Plant Families

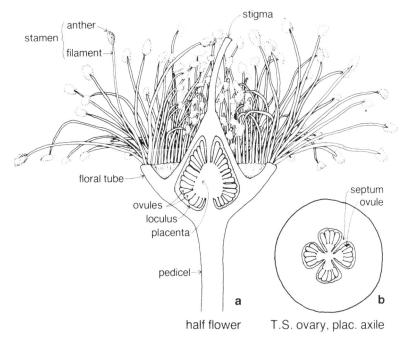

Fig. 81 *Eucalyptus camaldulensis* (river red gum) Myrtaceae
P operculum A∞ G ($\overline{4}$) floral tube present
Medium to tall, variable tree to 45 m. Trunk thick, main branches heavy, crown spreading; forest form more upright, with a smaller crown. The gum bark is variable in colour, grey, brown, or pinkish or white, and often patchy. Adult leaves lanceolate, usually 10–15 cm, sometimes to 25 cm long, dull green. Juvenile leaves opposite at first, then alternate; broad lanceolate. Flowers white, or creamy, 7–11 in axillary umbels. Fruit a capsule with exserted valves. Widespread along the river systems of all states except Tas. (a–b ×7)

MYRTACEAE

Fig. 82 *Eucalyptus globulus* ssp. *globulus* (Tasmanian or southern blue gum) Myrtaceae
Medium to very tall tree to 60 m. Trunk stout, branches relatively small, crown spreading. Bark of the gum-type, peeling in strips, so the trunk is streaked with dark and pale grey, blue-grey, cream and brown. Adult leaves alternate, lanceolate to 30 cm long, dark green. Juvenile leaves opposite, sessile, ovate, glaucous. Intermediate leaves very long, up to 60 cm. Flowers white, solitary, sessile, and axillary. Fruit a capsule. Widespread in Tas., restricted to the Otways and South Gippsland in southern Vic. Commonly grown in parks and gardens. Flowering winter to spring. (a–b ×0.5)

Plant Families

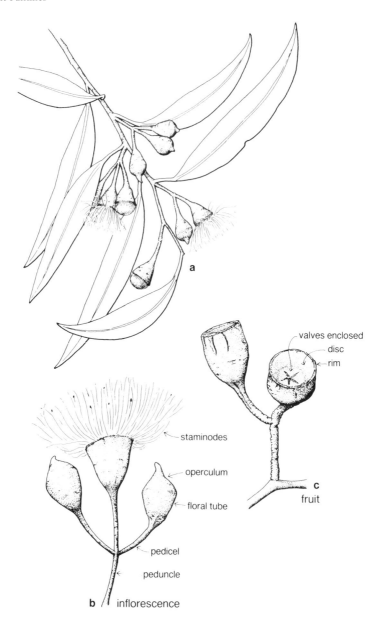

Fig. 83 *Eucalyptus leucoxylon* ssp. *megalocarpa* (large-fruited yellow gum) Myrtaceae
(a ×0.6, b–c ×1.2)

MYRTACEAE

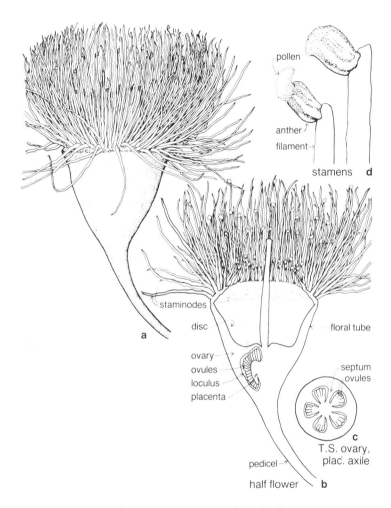

Fig. 84 *Eucalyptus leucoxylon* ssp. *megalocarpa* (large-fruited yellow gum) Myrtaceae

P operculum A∞ G($\overline{5}$) floral tube present

Small tree to about 20 m. Bark rough, box-like for about 2 m, then gum-type, whitish, cream or yellowish. Adult leaves lanceolate, dull green, to about 10 cm long. Juvenile leaves opposite, sessile, ovate, dull green. Flowers usually red or pink, and usually 3 in axillary umbels. The ring of tissue to which the stamens and staminodes are attached often persists in the fruiting stage, and may be called a staminophore or staminal ring. Fruit a capsule, valves enclosed. Mainly restricted to coastal calcareous regions of western Vic. and south-eastern SA. Extensively cultivated, often sold as *E. leucoxylon* 'Rosea'. Flowering mainly in autumn. (a–c ×2.5, d ×25)

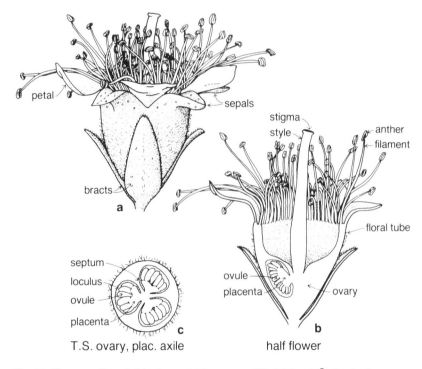

Fig. 85 *Kunzea ambigua* (white kunzea) Myrtaceae   K5 C5 A∞ G($\bar{3}$) floral tube present
Erect shrub to 3 m. Branchlets hairy. Leaves crowded, lanceolate-oblanceolate, up to 1 cm long, concave above. Flowers white or creamy in dense clusters on very short lateral branches. Each flower, at least when young, with bracts at the base; some authors state bracts are absent. Capsules, non-woody, with persistent calyx. Wilsons Promontory and coastal in eastern Vic., NSW and Tas. Flowering spring to summer. (a–c ×7)

MYRTACEAE

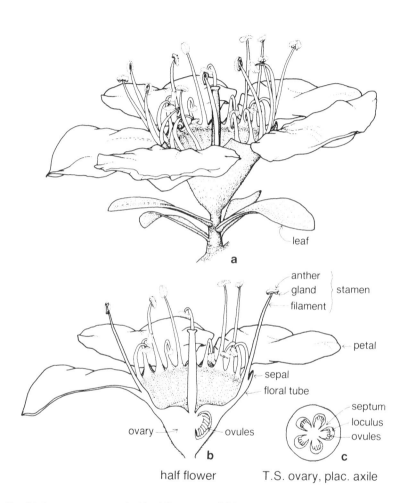

Fig. 86 *Leptospermum myrsinoides* (silky tea-tree) Myrtaceae

K5 C5 A∞ G($\overline{5}$) floral tube present

Shrub to 2 m tall. Leaves oblanceolate, up to 1 cm long, concave, dull green, and glabrous. Flowers numerous, white or pink, terminal on short lateral branchlets. Fruit a non-woody capsule. Widespread in heathlands and sandy forests, mainly coastal in Vic., SA and NSW. Flowering in spring. See also Pl. 3d. (a–c ×7)

Plant Families

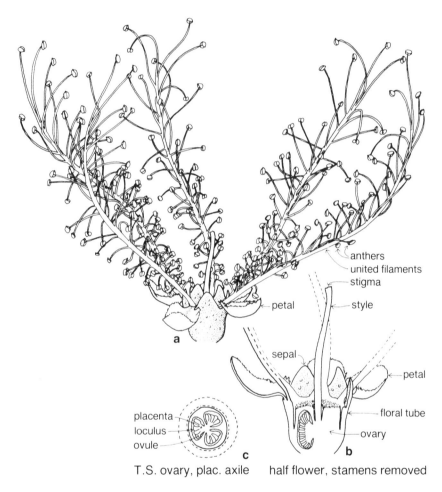

T.S. ovary, plac. axile   half flower, stamens removed

Fig. 87 *Melaleuca linariifolia* (snow-in-summer) Myrtaceae

K5 C5 A(∞) G($\overline{3}$) floral tube present

Shrub or small tree to 10 m. Leaves lanceolate and slightly concave, 2–3 cm long. Flowers white, borne in spikes near the shoot apex which continues to grow. Stamens united by filaments into 5 bundles (in some species the union is not so obvious; see also Pl. 6f). Fruit a capsule that persists unopened for some years. Coastal and damp localities in NSW and Qld. Widely planted as an ornamental and street tree. Flowering in summer. (a ×5, b–c ×7)

170

EPACRIDACEAE heath family

This medium-sized family of shrubs and small trees is predominantly Australian, with a few genera extending to South-east Asia, New Zealand and the Pacific Islands. Collectively the members of the family are known as heaths because they resemble the heaths and heathers of northern Europe, which belong to the family Ericaceae. The latter family has a worldwide distribution but is poorly represented in Australia by only four or five small genera including *Gaultheria* (wax-berry), found in the alpine areas of the south-east, and two native *Rhododendron* spp. in northern Qld. Among the well-known ornamentals from the Ericaceae are *Erica*, *Rhododendron* (including azaleas), and *Arbutus*. The flowers of the Ericaceae differ from the Epacridaceae in having twice as many stamens as petals, and 2-celled anthers. A recent proposal in fact combines these two into a larger Ericaceae.

The term heath is also applied to a plant community that is dominated by shrubs varying in height from 0.5 to 2.0 metres. These communities include many members of the Epacridaceae, and are found on shallow and/or poorly drained soils low in nutrients. A well-known member of the family is *Epacris impressa* (common heath, Fig. 89; Pl. 2d), the floral emblem of Victoria, and it is from this genus that the family name is derived.

Although *Epacris* is grown as an ornamental, other members of the family are not common in cultivation, apparently due to difficulties with propagation. The succulent fruits of some genera were eaten by the early settlers and Aboriginal people, who also valued the abundant nectar produced by many species.

Members of the family are illustrated in Figures 88–96 and Plate 2d–e.

FLORAL STRUCTURE

Flowers	Usually actinomorphic and bisexual, often small.
Calyx	Sepals 5, free. On the pedicel there are often additional bracts which may resemble the sepals and grade into them (Figs 89, 96).
Corolla	Petals 5 (sometimes 4 or 6), united. The corolla is often tubular, with 5 lobes at the top, that may be densely hairy, as in *Leucopogon* (beard-heath, Figs 90, 91). In the genus *Richea* (Fig. 93) the corolla remains closed, and as the anthers mature it splits transversely near the base, and the upper part (the operculum or cap) falls, leaving a cup-like rim.
Androecium	Stamens 5, often epipetalous, but hypogynous in *Dracophyllum*, *Prionotes*, *Richea*, *Sprengelia* and *Woollsia*. Anthers usually versatile, 1-celled at maturity, opening by a longitudinal slit.

Plant Families

Gynoecium   Carpels usually 4–5. Ovary superior with 1–10 loculi (mostly 4–5). Ovules 1–many per loculum. Placentation apical or axile. The style is simple and is either terminal (Fig. 91), or inserted in a depression in the top of the ovary (Fig. 89). A nectary disc or nectary glands are usually present at the base of the ovary. The nectary is absent in *Sprengelia* (swamp heath).
Fruit   A capsule or drupe.

Most of the Epacridaceae are small woody shrubs. The leaves are usually small, tough and pointed (often described as ericoid), sessile or shortly stalked and with the main veins parallel (Pl. 2e). A few genera, such as *Sprengelia* (Fig. 94), *Richea* and *Dracophyllum*, have leaves with broad sheathing bases. Some of the Tasmanian richeas resemble small palm trees, and specimens of *R. pandanifolia* up to 18 m high are found in the dense rainforests.

SPOTTING CHARACTERS

Woody shrubs. Leaves often small, tough and pointed (ericoid), with main veins parallel and more easily seen on the lower surface. Flowers actinomorphic, 5-partite. Sepals often grading into bracts of similar size and texture. Corolla often tubular.

ROUTES THROUGH KEYS

Routes through several keys are given here for members of the family Epacridaceae illustrated in this book, although not all species are included in all keys. Texts are cited by author's name followed by the bibliography number.

BEADLE *et al.* (18)
Key to families, p. 99.
A-dicots
-*A-*B-*C-*D-*E-L-*M-N-*O- group 11 -*A-*F-G- Epacridaceae
-*A-*B-*C-*D-*E-L-*M-*N- group 12 -*A-*C-*D-*E-*H-*I-*J
 - Epacridaceae

Use the alternative key to genera, p. 401 (reproduced in Ch. 7, as Key 2).
A-B- *Sprengelia* -A-B- *S. incarnata*
*A-C-*D-E- *Leucopogon* -A-*B-C- *L. virgatus*
                          *A-*D-*E-*I-*L-N-*O- *L. ericoides*
*A-*C-F-*G-*H-I- *Woollsia* - *W. pungens*
*A-*C-F-*G-*H-*I-J- *Epacris*
*A-*C-*F-*K-*L-*M-*N- *Lissanthe* -A- *L. strigosa*

EPACRIDACEAE

BLACK (24)
Key to families, p. 23 (see note about Black's key to families, in the section on Proteaceae). Alternative leads in a couplet are here designated A1, A2, A3; B1, B2, etc.
A2-B2-C2- subclass 2 (p. 30) -A1-B1- Epacridaceae

Key to genera, p. 665.
A1-B1-C2-D3-E1- *Lissanthe* - *L. strigosa*
A1-B1-C2-D3-E2- *Leucopogon* -A1-B2-C2-D2-E2- *L. virgatus*
                                            -A2-F1-G1- *L. ericoides*
A2- *Epacris* - *E. impressa*
A2- *Sprengelia* - *S. incarnata*

CAROLIN AND TINDALE (42)
First key to families, p. 118.
1-Dicots-1-2-3-4-5-12-13-15- Group 12 -1-5-6-7-16-17-19- Epacridaceae
1-Dicots-1-2-3-4-5-12-13-15-16- Group 14 -1-8-9-10- Epacridaceae

Key to genera, p. 460.
1-2-3-4-6-7-8- *Leucopogon* -1-2-3- *L. virgatus*
                          -1-5-6-10-13-17-18-19- *L. ericoides*
1-2-3-4-6-7-8-9-10- *Lissanthe* -1-*L. strigosa*
1-11-12- *Sprengelia* -1-2- *S. incarnata*
1-11-13-14- *Epacris*

WILLIS (233)
Key to families, vol. 2, p. 1.
1-4-6-7-8-11-88(D)-89-92-93-94-97-105-111-116-117-118- Epacridaceae

Key to genera p. 495
1-2-3- *Epacris* -1- *E. impressa*
1-2-3-4- *Sprengelia* - *S. incarnata*
1-2-5-6-7-8-9-10-12-13- *Lissanthe* - *L. strigosa*
1-2-5-6-7-8-9-10-12-13-14- *Leucopogon* -1-2-3-4- *L. virgatus*
                                 -1-5-6-10-11-12-13- *L. ericoides*

Plant Families

Fig. 88 *Epacris impressa* (common heath) Epacridaceae (×0.7)

EPACRIDACEAE

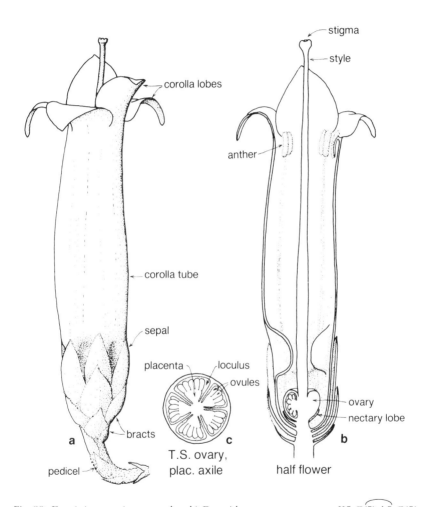

Fig. 89 *Epacris impressa* (common heath) Epacridaceae   K5 C(5) A5 G(5)
Erect to spreading shrub to about 1 m high. Leaves lanceolate, to 1.5 cm long, pungent, midrib prominent on the underside. Flowers white, pink or red, axillary. Fruit a capsule, opening by 5 valves. Widespread in heaths, and drier forests with a heathy understorey, in Vic., Tas., NSW and SA. Flowering autumn to spring. See also Pl. 2d. (a–b ×7, c ×12)

Plant Families

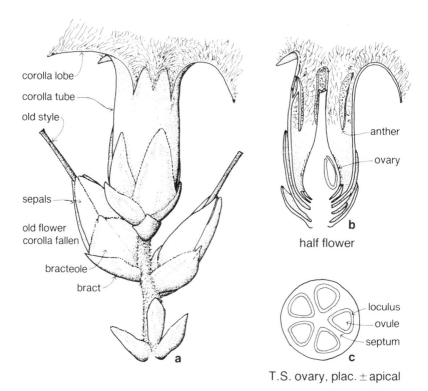

Fig. 90 *Leucopogon ericoides* (pink beard-heath) Epacridaceae K5 C(5) A5 G(5)
Slender, wiry, small shrub to 2 m. Branches pubescent. Leaves oblong, less than 1 cm long, mucronate, margins recurved. Flowers white to pink, 2–4 in short axillary spikes. Each flower subtended by a bract and 2 bracteoles. Fruit a drupe. Widespread in heathlands, Vic., Tas., NSW, SA and Qld. Flowering in spring. (a–b ×15, c ×30)

EPACRIDACEAE

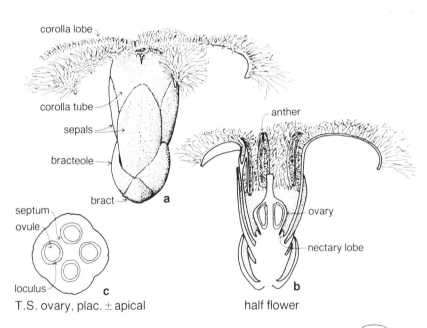

Fig. 91 *Leucopogon virgatus* (common beard-heath) Epacridaceae   K5 C(5) A5 G(4)
Wiry small shrub to 1 m. Leaves to 1.5 cm long, lanceolate, incurved, sharply pointed, lower surface with conspicuous parallel veins. Flowers white in short, dense, axillary spikes, each flower subtended by a bract and 2 bracteoles. Number of loculi in the ovary variable, usually 4 or 5. Anthers with sterile tips. Fruit a drupe. Widespread in heathlands, Vic., Tas., NSW, SA and Qld. Flowering in spring. (a–b ×15, c ×40)

Plant Families

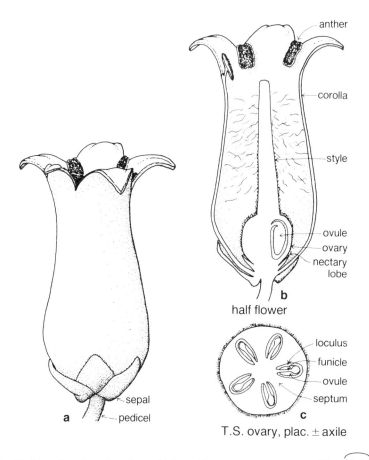

Fig. 92 *Lissanthe strigosa* (peach heath) Epacridaceae K5 C(5) A5 G(5)
Small rigid shrub, ± 0.5 m high. Leaves linear, pungent, to 1.5 cm long, margins recurved. Flowers pale pink in small axillary clusters. Fruit a drupe. Widespread in all states except WA. Flowering late winter to spring. (a–b ×12, c ×25)

EPACRIDACEAE

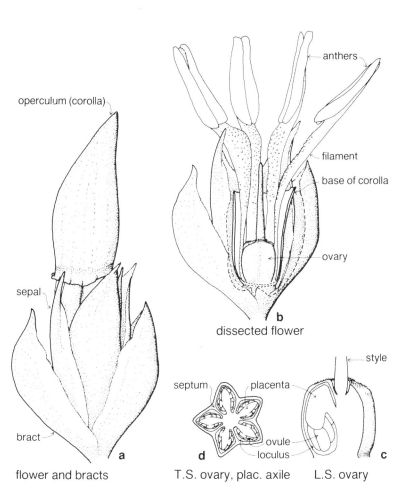

Fig. 93 *Richea procera* (richea) Epacridaceae   K5 C(5) A5 G(5)
Sparingly branched shrub to 3 m. Leaves imbricate, with sheathing bases, recurved, ovate-lanceolate and tapering to a point, 1–3.5 cm long. Flowers creamy yellow, arranged in short spikes or heads terminating main or lateral shoots. The corolla splits transversely, and the upper part is shed as a cap or operculum. Filaments thickened and papillose in the upper part. Endemic in Tas., where occasional but locally common. Flowering late spring. (a–b ×10, c–d ×20)

Plant Families

Fig. 94 *Sprengelia incarnata* (swamp heath) Epacridaceae (×0.7)

EPACRIDACEAE

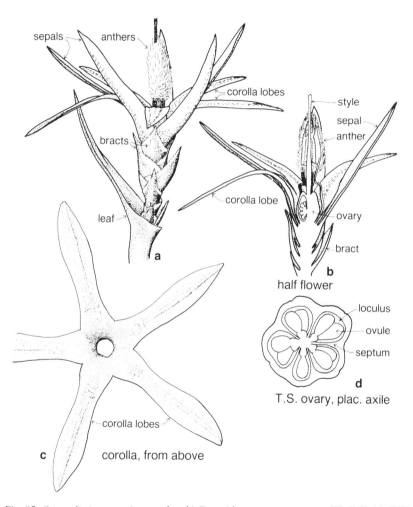

Fig. 95 *Sprengelia incarnata* (swamp heath) Epacridaceae K5 C(5) A5 G(<u>5</u>)
Shrub to 2 m, with rigid erect stems. Leaves with sheathing bases, to 2 cm long, tapering to a sharp point. Flowers pink in leafy terminal clusters. Fruit a capsule. Wet heaths and swamps in southern Vic., SA, Tas. and NSW. Flowering in spring. (a–c ×7, d ×25)

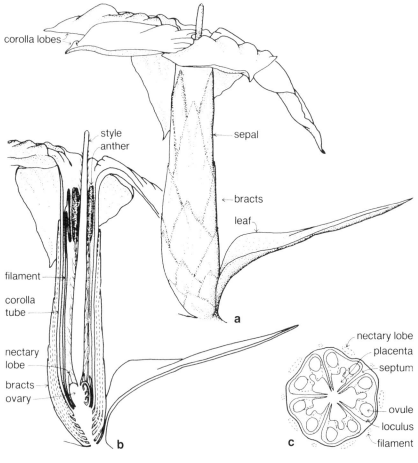

Fig. 96 *Woollsia pungens* (woollsia) Epacridaceae   K5 C(5) A5 G(5̲)

a flower with enveloping bracts and subtending leaf; **b** half flower—the dotted lines indicate the limits of the bracts and sepals, to distinguish them from the corolla tube; **c** T.S. ovary, placentation axile (a–b ×7, c ×25)

Erect shrub to 2 m high. Leaves crowded, sessile, ovate, acuminate, about 1 cm long. Flowers white or reddish, sessile and axillary. Bracts around the base of the flower grade into the 5 sepals. The anthers are borne on very fine filaments that are free of, but may appear attached to, the corolla tube. Fruit a capsule. Common on coastal heaths and in dry forests, NSW and Qld. Occasionally cultivated. Flowering in winter and early spring to

SOLANACEAE nightshades

The Solanaceae is a large family, predominantly of the southern hemisphere, and particularly in South and Central America. The family contains approximately 84 genera, of which 24 occur in Australia.

Alkaloids are found in the foliage, flowers and immature fruits of most genera. Many are toxins, and some are important in the pharmaceutical industry. Solandine, used in contraceptive pills, is extracted from *Solanum aviculare* and *S. laciniatum* (kangaroo apples, Pl. 8c). Aboriginal people chewed the leaves of native members of *Nicotiana* (Pl. 8i), and the American species *N. tabaccum* is the source of commercial tobacco. The leaves of *Duboisia hopwoodii* (pituri, Pl. 8h) contain nicotine and nor-nicotine, and were also chewed by Aboriginal people as well as being used as a source of animal poison.

Family members cultivated for food include *Capsicum* spp. (peppers and chillies), *Lycopersicon esculentum* (tomato, Pl. 8e) and *Solanum tuberosum* (potato). Potatoes are underground storage organs produced on the ends of stolons arising from the plant axis. Well-known ornamentals are *Cestrum* spp. (Pl. 8a), *Brugmansia* spp. (angel's trumpet, Pl. 8g), *Nicotiana* spp., *Petunia* hybrids, *Physalis* spp., and *S. jasminoides* (potato vine), *S. aviculare* and *S. laciniatum* (kangaroo apples, Pl. 8c).

A number of introduced species have become troublesome weeds including *Datura* spp. (thornapples), some *Nicotiana* and *Physalis* spp. (Pl. 2c), *Lycium ferocissimum* (boxthorn), *Solanum nigrum* (black nightshade, Pl. 8b), and *Salpichroa origanifolia* (pampas lily of the valley, Pl. 8f).

Members of the family are illustrated in Plates 2c, 2f and 8. For descriptions and keys to Australian species see *Flora of Australia* vol. 29 (bibliography no. 94).

FLORAL STRUCTURE

Flowers — Actinomorphic or slightly zygomorphic, mostly bisexual. Inflorescence various, often terminal, or flowers solitary.

Calyx — Sepals 5, united. Calyx tubular to campanulate, mostly 5-lobed (occasionally 3–9 lobed), persistent and sometimes developing to enclose the fruit (*Physalis*, Pl. 2c).

Corolla — Petals 5 (3–9), united. Corolla may be tubular, campanulate, urceolate or funnel-shaped (Fig. 2g, j, k). The limb is often rotate or stellate (Fig. 2h).

Androecium — Stamens usually 5 or 4, epipetalous and alternating with the corolla lobes. Anthers sometimes coherent, dehiscing by vertical slits or apical pores (Pl. 8c).

Gynoecium — Carpels 2–5, united. Ovary superior, with a nectary disc usually present (Pls 2f, 8f). Placentation axile, the ovules usually numerous (Pl. 2f).

Fruit — A berry, sometimes rather dry, or a capsule.

Plant Families

The family includes herbs, shrubs, small trees, and some climbers. Stems are glabrous or hairy, or with prickles. Leaves are alternate, occasionally almost opposite, simple, entire, deeply lobed or pinnate, exstipulate, and sometimes prickly.

SPOTTING CHARACTERS

Flowers usually actinomorphic, with 5 united sepals, and 5 united petals. Corolla often rotate (Fig. 2h, Pl. 8b). Stamens 5, epipetalous. Fruit a berry or capsule, with many seeds.

LAMIACEAE mint family, mint-bushes

(LABIATAE)

The family, found throughout the world, is particularly well represented in the Mediterranean region, and about 38 genera are native in Australia. The older name, Labiatae, is derived from the character of the corolla, in which one or more petals form a distinct lip. The alternative name is derived from the Mediterranean genus *Lamium*.

The family includes many aromatic culinary herbs, such as *Origanum* (marjoram), *Mentha* (mint), *Thymus* (thyme), *Rosmarinus* (rosemary) and *Salvia* (sage), all of which were early introductions to Australia. *Lavandula dentata* (lavender) is cultivated in Tasmania and Victoria for the commercial extraction of the oil, which is used in perfumes. The many introduced genera grown as ornamentals include *Coleus* (painted nettle), *Leonotis* (lion's ear), *Nepeta* (cat mint), *Lavandula*, *Salvia* (Fig. 98) and *Stachys* (lamb's ears). Two well-known native genera widely cultivated are *Westringia* and *Prostanthera* (mint-bush, Fig. 97). Several European genera are declared noxious weeds, including *Marrubium vulgare* (horehound), *Lavandula stoechas* (topped or Spanish lavender) and *Lamium amplexicaule* (henbit dead-nettle).

For further reading see bibliography numbers 2, 3 and 124.

FLORAL STRUCTURE

Flowers	Usually zygomorphic, bisexual. Usually arranged in axillary cymose clusters (and may appear whorled around the stem), rarely solitary.
Calyx	Sepals 5, united. Calyx often 2-lipped.
Corolla	Petals 4–5, united. Corolla tubular but usually deeply lobed, often 2-lipped.
Androecium	Stamens 4 or 2, epipetalous, sometimes arranged in 2 pairs of unequal length.
Gynoecium	Carpels 2. The superior ovary is deeply divided, with 4 loculi. Style gynobasic, stigma usually bifid (forked). Ovules 1 per loculus. Disc usually present.
Fruit	At maturity the ovary splits into 4 achene-like nutlets.

The majority of mints are aromatic shrubs or herbs. The stems are often quadrangular with simple, opposite, or occasionally whorled leaves. There are no stipules.

SPOTTING CHARACTERS

Flowers zygomorphic, stamens 2 or 4. Stems square (in T.S.) and leaves opposite and aromatic.

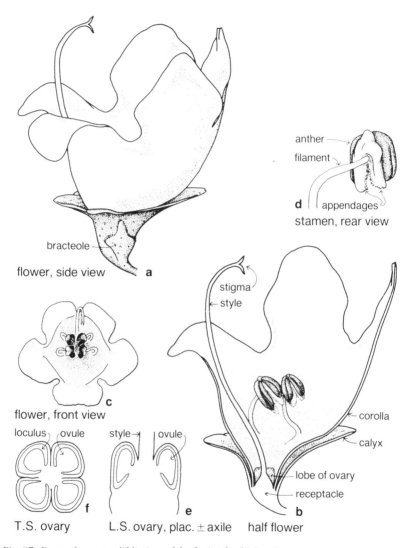

Fig. 97 *Prostanthera rotundifolia* (round-leaf mint-bush) Lamiaceae

K(5) C(5) A4 G(loculi 4)

Bushy shrub to 4 m. Leaves opposite, aromatic, more or less orbicular, about 1 cm long. Flowers mauve-purple, profuse in bracteate racemes. Calyx with bracteoles and dotted with oil glands. Anthers with appendages. On dry hills and along watercourses in Vic., Tas., SA and NSW. Many members of the genus are cultivated. Flowering in late spring. (a–b ×7, c ×2.5, d ×10, e–f ×20)

LAMIACEAE

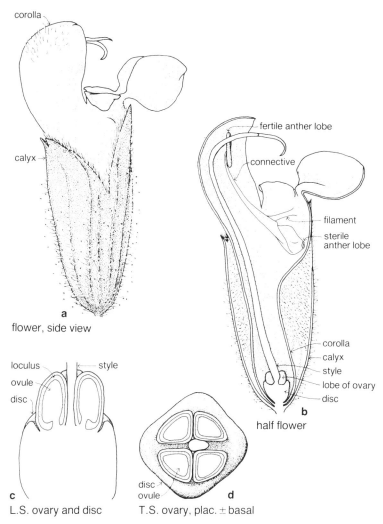

Fig. 98 *Salvia verbenaca* (wild sage) Lamiaceae K(5) C(5) A2 G(loculi 4)
Hairy herbaceous perennial to 60 cm high. Leaves opposite, ovate-oblong, wrinkled on the surface. Margins crenate, often lobed. Flowers purplish, rarely white, in apparent whorls of about 6, in the axils of orbicular bracts. Both calyx and corolla 2-lipped. Each anther has a very broad connective that separates the fertile lobe from the sterile one. The sterile lobe and connective act as a lever to move the fertile lobe. A foraging insect pushing against the lever may deposit pollen on its back. Introduced from Europe, a weed in all southern states. Flowering in summer. Many *Salvia* spp. are cultivated. (a–b ×6, c–d ×24)

187

MYOPORACEAE myoporums and emu-bushes

Myoporaceae, a small family of only three woody genera, is most developed in Australia, but is also found in the West Indies, some Indian Ocean islands and the south-west Pacific. In Australia, *Myoporum* and *Eremophila* are widespread; the other genus *Bontia*, is found only in the West Indies.

The largest genus, *Eremophila* (emu-bush, Pl. 9c–f) is an Australian endemic with about 180 species, most of which grow in the arid zone. Many are in cultivation and make attractive garden plants.

There are about 31 species of *Myoporum* with half of these found mostly in southern and temperate parts of Australia. *M. insulare* (boobialla, Pl. 9b) has been widely planted in windbreaks in South Australia and Victoria, and *M. floribundum* and *M. parvifolium* (creeping boobialla) make attractive ornamentals. *M. platycarpum* (false sandlewood) is locally common in low woodlands in the arid zone.

For further reading see bibliography no. 208.

FLORAL STRUCTURE

Flowers	More or less actinomorphic in *Myoporum*, zygomorphic in *Eremophila*. Bisexual. Flowers axillary, occurring singly or in groups of 2–11.
Calyx	Sepals 5, occasionally 4, mostly free, sometimes united at the base to form a tube, often enlarging after flowering.
Corolla	In *Myoporum*, petals 5, united at the base with a short tube and spreading lobes, or campanulate. In *Eremophila*, petals 5, united. Corolla tubular or campanulate, the upper lip often of 4 or 2 lobes, lower lip of 1 or 3 lobes (Pl. 9c–f).
Androecium	In *Myoporum*, stamens 4, exserted. Filaments mostly straight. In *Eremophila*, stamens 4 or 5, exserted (Pl. 9c–d) or included (Pl. 9e–f) within the corolla. Filaments straight or curved.
Gynoecium	Carpels 2, united, ovary superior. After fertilisation the ovary may become further divided due to outgrowths from the placentas. Placentation axile.
Fruit	In *Myoporum* usually a fleshy drupe. In *Eremophila* a fleshy drupe, or dry.

Most members of the family are shrubs or small trees. Leaves are alternate, rarely opposite or whorled, simple, mostly sessile and exstipulate. The plant tissues contain large embedded resinous ducts, sometimes occurring as surface tubercles or resin-secreting glandular papillae.

GOODENIACEAE

This family is mostly confined to the southern hemisphere, with all of the 11 genera found in Australia, a few species extending to South-east Asia. The type genus *Goodenia* is the largest and is widespread in south-eastern Australia. The Western Australian genus *Lechenaultia* is the most widely cultivated member of the family.

Aboriginal people utilised a few species, such as *Scaevola spinescens*, the drupes of which were much prized by Aboriginal people in the Flinders Ranges.

For descriptions and keys to Australian species see *Flora of Australia* vol. 35 (bibliography no. 94).

FLORAL STRUCTURE

Flowers	Zygomorphic and bisexual.
Calyx	Sepals 5, free, or united at the base.
Corolla	Petals 5, united below, but the corolla split almost to the base on one side, often 2-lipped, sometimes spurred as in *Velleia paradoxa* (spur velleia). The corolla lobes are sometimes spread like a fan as in *Scaevola* (fan-flower, Pl. 3c). The lobes are often described as winged, because a central band along the midrib differs in colour or texture from the thinner marginal wings that are folded under in bud (Fig. 99).
Androecium	Stamens 5, free or with anthers joined in a tube around the style. Stamens sometimes epipetalous at the base of the corolla tube.
Gynoecium	Carpels 4, united, but not apparent in T.S. ovary as there are only 1 or 2 loculi. Sometimes the septum is incomplete at the top of the ovary, in which case the ovary is incompletely 2-celled. Ovules 1–many. Ovary usually inferior to semi-inferior, sometimes superior. At the top of the style is a 2-lipped or cup-shaped structure, usually with hairy margins, called an **indusium** (Fig. 99; Pl. 3c). Before the flower opens the pollen is shed into the indusium which is carried upwards by the lengthening style. Later, the pollen is pushed out by the growth of the stigma, which is usually bilobed.
Fruit	A capsule, drupe or nut.

Members of the family are annual or perennial herbs or low-growing shrubs. The leaves are usually alternate or radical. There are no stipules.

SPOTTING CHARACTERS

Flowers usually zygomorphic. Corolla lobes winged. The indusium at the end of the style is very characteristic of the family.

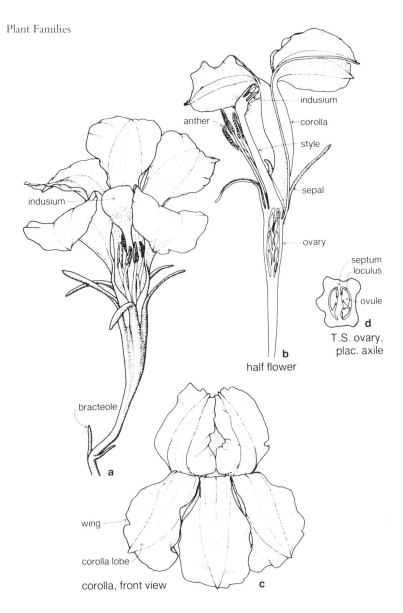

Fig. 99 *Goodenia ovata* (hop goodenia) Goodeniaceae   K5 C(5) A5 G(loculi 2)
Spreading shrub to 2 m high. Leaves ovate, thin, 2–5 cm long, with finely toothed margins. Flowers yellow, 1–3 together in the leaf axils. Corolla lobes conspicuously winged. Indusium present on the style. The septum often incompletely divides the loculi. Fruit a cylindrical capsule about 1 cm long. Widespread near the coast and in mountain forests, Vic., SA, Tas. and NSW. Flowering mainly spring and summer. (a–c ×4, d ×8)

BRUNONIACEAE

This family, of a single genus *Brunonia* (blue pincushion, Pl. 6c) is sometimes included in the Goodeniaceae. The only species, *B. australis*, is widely spread throughout Australia, often in grasslands and drier forests. It is an herbaceous perennial, with a basal rosette of hairy grey-green oblanceolate leaves, and the inflorescence is a head, borne on a slender stalk about 20–25 cm tall. An involucre of green bracts surrounds numerous, small, sessile blue flowers. The style, which protrudes from the flower, is terminated by a tiny yellow indusium (cf. *Goodenia*, Fig. 99). For a botanical description see *Flora of Australia* vol. 35 (bibliography no. 94).

STYLIDIACEAE trigger-plants

While predominantly Australian, this family has a few representatives in Southeast Asia, New Zealand and the tip of South America.

The largest genus, *Stylidium* (Fig. 100, Pl. 9h) is widespread in Australia. Most members are annual or perennial herbs, with a cluster of basal leaves from which the stalk or scape bearing the inflorescence arises. The flowers are bisexual and zygomorphic. There are 5 sepals and the calyx is often 2-lipped. The corolla, of 5 united, white to deep pink petals, is deeply lobed with a short tube. One lobe is comparatively short, a different shape, often reflexed, and is called the labellum. At the throat of the corolla there may be 6 or 8 prominent erect appendages (Pl. 9h). The filaments of the 2 stamens are united with the style to form a column. The anthers are attached at the top of the column, with the stigma between them. The ovary is inferior, 1- to 2-celled, with many ovules attached to a central placenta. The fruit is a capsule. The calyx, the outer casing of the ovary, and the scape are often glandular hairy.

The column of *Stylidium* is sensitive to touch, and at first is reflexed over the labellum. When the base of the column is touched by an insect collecting nectar, the column springs up and the pollen is showered over the back of the insect. When the pollen has dispersed, the column again reflexes and the stigma becomes receptive. If the next insect to trigger the column has pollen on its back, this may then be transferred to the receptive stigma.

For further reading, see number 81 in the bibliography.

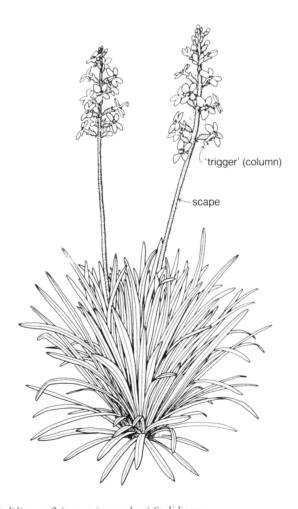

Fig. 100 *Stylidium* sp. 2 (grass trigger-plant) Stylidiaceae
This species is at present without a name and is designated as 'sp. 2' in the *Flora of Victoria* vol. 4, p. 587. It was formerly included within *S. graminifolium*. See also Pl. 9h. (×0.4)

ASTERACEAE daisy family

(COMPOSITAE)

The Asteraceae, often known as Compositae, is a very large family of worldwide distribution, and includes all the daisies, thistles and everlastings. Over 200 genera and nearly 1000 species are found in Australia. The name is derived from the genus *Aster*, which includes familiar garden plants such as the Michaelmas and Easter daisies. The family is readily recognised because the apparent 'flowers' are condensed inflorescences of many small flowers clustered together. It is this composite nature of the 'flower' that gave rise to the older name Compositae, which is still widely used.

The family is a highly successful one due to its adaptability and high reproductive rate. Many introduced species are proclaimed weeds; local examples include the thistles, capeweed, boneseed and ragwort. Members of the family are not planted in pastures but some are important as food plants, such as *Lactuca sativa* (lettuce), *Cichorium endiva* (endive) and *Cynara scolymus* (artichoke, Pl. 10j). Safflower oil is extracted from the seeds of *Carthamus*, sunflower oil from the seeds of *Helianthus*, and the culinary herb tarragon is *Artemisia dracunculus*. Many species are grown as ornamentals, including *Aster*, *Calendula* (English marigold), *Chrysanthemum*, *Dahlia*, *Helichrysum* and *Zinnia*.

For further reading on Australian everlastings, see bibliography no. 10.

FLORAL STRUCTURE

Flowers   Regular or zygomorphic, usually bisexual or female. Arranged in heads (capitula), and usually referred to as **florets** (Fig. 102a–b, etc.).
Calyx   If present, usually represented by scales, hairs, bristles or awns, and known as a **pappus**.
Corolla   Petals usually 5, sometimes 3 or 4, united.
Androecium   Stamens usually 5 and epipetalous, united by their anthers, which form a tube around the style (Fig. 102d).
Gynoecium   Carpels 2 (indicated by 2 style-arms), united. Ovary inferior, unilocular with 1 basal ovule.
Fruit   Usually dry and indehiscent and then commonly called an **achene** (but technically a **cypsela**—achenes are derived from superior ovaries). Occasionally a drupe, as in the weed *Chrysanthemoides monilifera* (boneseed, Pl. 10b).

Within the family there is considerable variation in floral structure. The following notes amplify the brief summary given above. Most of the features discussed are illustrated in Figures 101–12 and Plate 10.

The inflorescence is called a **capitulum** or **head**, and the individual flowers are known as **florets**. The florets, which are sessile (stalkless), are borne on the **receptacle**, the expanded end of the inflorescence stalk. The receptacle is usually more or less flat, but in some genera is elongated to form a club-shaped structure. Its surface is either smooth or pitted, and sometimes there are hairs or small scales at the base of each floret (Pl. 10a). If hairs or scales are absent, the receptacle is said to be naked.

Each head is surrounded by a series of bracts, which are usually referred to collectively as the **involucre**, and the component parts as **involucral bracts**.

Two types of floret, tubular and ligulate, are most commonly observed.

Tubular florets, also called **disc** florets, (Figs 102b, 103d) have a tubular, actinomorphic corolla of 5 (sometimes 4) united petals with 5 (4) lobes at the top. Filiform florets are a form of disc floret in which the corolla tube is very slender, and these are usually unisexual.

Ligulate florets, also called **ray** florets (Figs 102a, 103c, etc.; Pl. 10g), have a short corolla tube at the base, which extends on one side into a flat strap-shaped part called the ligule or ray. There are either 3 or 5 lobes or teeth at the end of the ligule, said to represent the number of petals.

At anthesis the pollen is shed to the inside of the anther tube and pushed out of the corolla by the upward growth of the style, to be gathered by an insect, or otherwise dispersed. Finally the style-arms separate and the stigmatic surfaces become receptive.

Depending on the arrangement of the floret types, three kinds of head are common: radiate, discoid, and ligulate.

In the **radiate head** (Figs 101, 103; Pl. 10a–e) both tubular and ligulate florets are present. The tubular florets are usually bisexual, occupy the centre of the receptacle, and may be referred to collectively as the disc. The ligulate florets are arranged in one or more rows around the edge of the receptacle and collectively may be called the ray. They commonly have a 3-toothed ligule and are often either female or neuter.

The **discoid head** (Figs 104–8) has only tubular florets, which may be all bisexual, or some female or neuter florets may be found in the outer part of the head (next to the involucral bracts). In addition, the corollas of these florets may be filiform or absent.

The **ligulate head** (Figs 109–10; Pl. 10f, h) contains ligulate florets only, which are nearly always all bisexual, but occasionally may be all male or all female. The florets usually have a 5-toothed ligule (Pl. 10g).

Florets in the head vary in number with species and they may be of similar or different sex. The head is said to be homogamous if all florets are of the same sex and heterogamous if a mixture of sexes is present. The most common combination is bisexual and female, with the bisexual florets in the centre and one or two rows of female florets around the outside of the head (Fig. 101b). As the head is a racemose inflorescence, the youngest florets are in the centre and are the last to mature (Pl. 10d).

Heads may be solitary or arranged in similar ways to other flowers, that is, in racemes or panicles etc. Some species have many small heads closely packed together at the end of a single stalk resulting in a **compound head** (Fig. 111; Pl. 10i). Each small head within the compound head will have a **partial involucre** and is called a **partial head** (Fig. 112). Bracts at the base of a compound head are referred to as the **general involucre**.

As the family is so large, fine detail is often required for identification of species. The following paragraphs emphasise some further features of flower structure used in classification.

The shape of the base of the **anther lobes** and the presence or absence of apical and/or basal appendages is of importance. If a short basal appendage is present, the anther lobes are described as acute; if the appendage is longer, it is called a **tail** and the lobes are said to be **tailed** or **caudate** (Fig. 108c). As the anthers start to wither after the pollen is shed, examination of the lobes for appendages is best carried out on buds.

Style-arms vary in form. They may be flattened or rounded in cross-section, and their apices may be acute, obtuse or truncate. The stigmatic areas are often papillose and may cover the inner surface of each arm or be more restricted.

The **pappus** may be of hairs, bristles, scales or awns. The bristles are said to be simple if they are smooth, barbellate if they have short rough appendages, and plumose if they bear long fine hairs. Scales are flat, bract-like structures, which may be glabrous or pubescent, and are said to be fringed if the hairs are only marginal.

Involucral bracts are usually numerous, in one or more rows, and may be free from one another or united. They vary considerably in structure and texture and may be herbaceous (leaf-like and green), leathery, scaly, stiff or spiny. The margins of herbaceous bracts are often dry, colourless and membranous, and are then said to be **scarious**.

The **fruit** is usually dry, 1-seeded and indehiscent and its correct name is a **cypsela**. However, it is commonly referred to as an **achene** (the term 'achene' is used properly to describe a dry, 1-seeded fruit derived from a superior ovary). If a pappus is present in the flower, it usually persists at the fruiting stage and is often an aid to fruit dispersal (Figs 102c, 109c). A dry, hairy, persistent pappus enables the fruit to be blown away by the wind and many thistles are dispersed in this way. Sometimes the tissue between the pappus and the ovary grows as the fruit develops, and the pappus is then connected to the top of the fruit by a small stalk called the **beak** (Fig. 109c). The whole structure is often then referred to as a **beaked achene**.

Young vegetative parts of some species, when damaged, exude a white sticky juice known as **latex**. This is typical of the tribe Cichorieae, to which dandelion, lettuce and some of the thistles belong.

Most members of the Asteraceae are annual, biennial or perennial herbs. Some are exceedingly small while others, like the sunflower, are large with spectacular inflorescences. A few genera are woody shrubs or small trees. Leaf characters vary enormously, but there are no stipules.

SELECTED EXAMPLES

The illustrations (Figs 101–12) show examples of species with radiate, discoid, ligulifloral and compound heads. Floral formulae have been omitted from individual captions. They all take the form K pappus $C\overline{(5)\ A(5)}\ G(\overline{2})$, except where corollas are 4-partite or florets unisexual.

ASTERACEAE

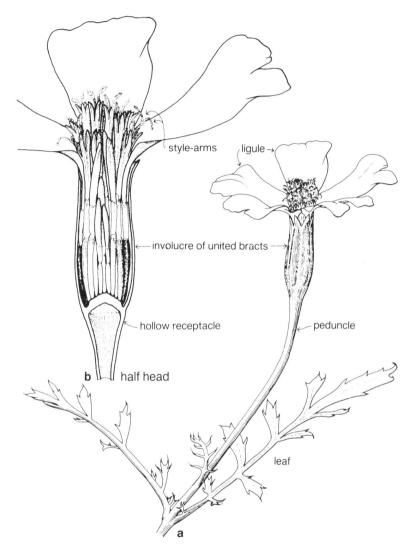

Fig. 101 *Tagetes* 'Cinnabar' (marigold) Asteraceae
Herbaceous annual to 30 cm high. Leaves mostly opposite, pinnatisect, up to 10 cm or more long, strongly aromatic. Flower heads radiate, surrounded by one row of united involucral bracts. Ligulate florets female, deep red with a yellow border; tubular florets bisexual, yellow. Pappus of white scales. Ovary darkening with development, becoming almost black in fruit. One of many cultivated marigolds, most of which are double and do not show the basic structure. Flowering summer to autumn. The genus is native to South and Central America and adjacent North America. (a ×1.5, b ×3)

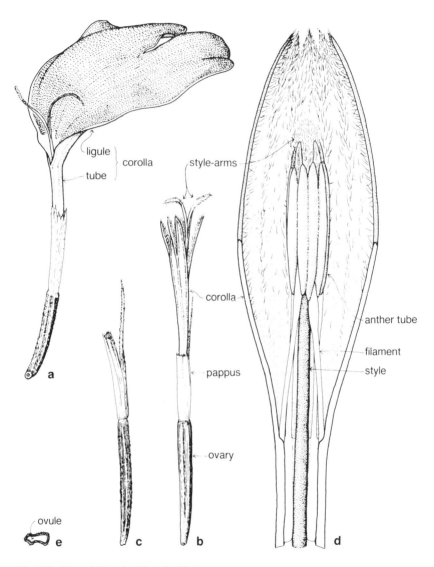

Fig. 102 *Tagetes* 'Cinnabar' (marigold) Asteraceae
a ligulate floret; **b** tubular floret; **c** fruit; **d** young tubular floret—upper part of corolla opened out to show anther tube around style and emerging style-arms; **e** T.S. mature ovary—the ovule completely fills the loculus, placentation basal (a–c ×3, d ×12, e ×6)

ASTERACEAE

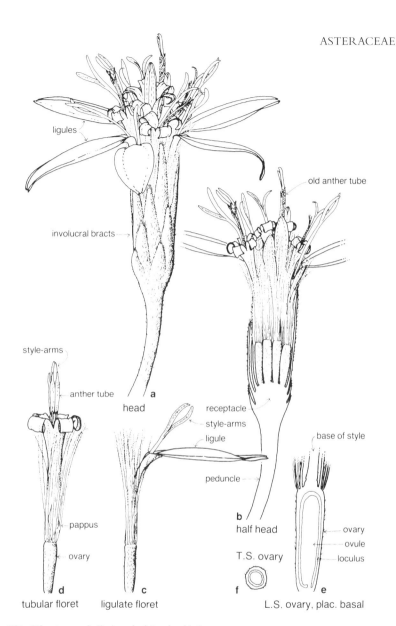

Fig. 103 *Olearia argophylla* (musk daisy-bush) Asteraceae
Small tree to 10 m high. Leaves broad-lanceolate to ovate, to 15 cm long, with a musky odour, margins slightly toothed, upper surface green, lower surface covered with silvery hairs. Flower heads radiate, creamy white, numerous in large panicles. Pappus of numerous capillary bristles. Common in cool gullies and sheltered forests of Vic., Tas. and NSW. Flowering in spring and summer. (a–d ×6, e–f ×12)

Plant Families

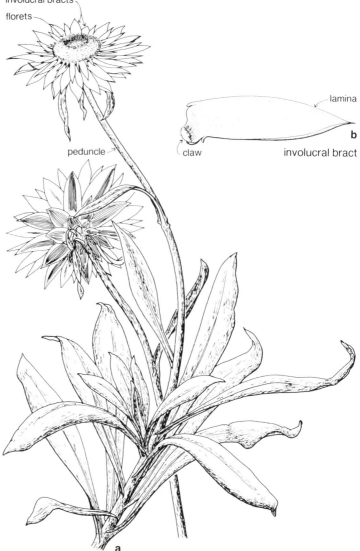

Fig. 104 *Xerochrysum bracteatum* 'Dargan Hill Monarch' (everlasting) Asteraceae
Spreading perennial to 50 cm. Leaves lanceolate, silvery grey, hairy. Flower heads golden yellow, large, up to 7 cm diameter. Florets all tubular, surrounded by spreading papery involucral bracts. These resemble ligulate florets and remain showy when the central florets have matured and fallen, hence the name 'everlasting'. Previously in the genus *Helichrysum*, and later transferred to *Bracteantha* but *Xerochrysum* was found to have priority. The species is variable and widespread, this form native to Qld (Cunninghams Gap). Flowering spring and summer. A selected cultivar, commonly grown, the cultivar name written in single quotation marks. (a ×0.6, b ×2)

ASTERACEAE

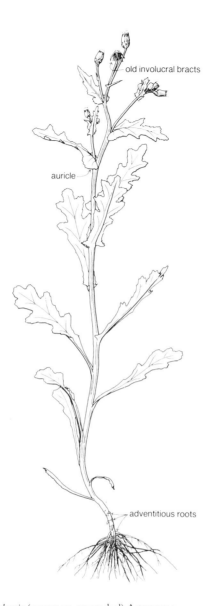

Fig. 105 *Senecio vulgaris* (common groundsel) Asteraceae
Erect annual herb to 30 cm high, stem glabrous or with scattered white hairs. Leaves variably pinnatifid with widely spaced and sinuate-toothed lobes. Upper leaves tend to be stem-clasping (amplexicaul). Florets all tubular, yellow. Involucral bracts usually described as being in one row in the genus, but in this case with a number of additional bracts at the base of the involucre, often conspicuously tipped black. Anther bases obtuse. Pappus of numerous capillary bristles. Widespread naturalised weed, common in gardens, introduced from Europe. Flowering mainly winter to spring. (×0.5)

Plant Families

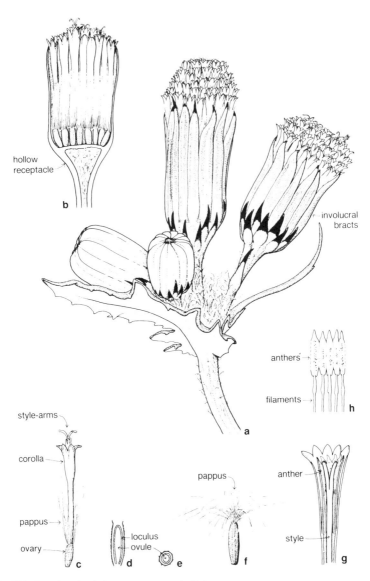

Fig. 106 *Senecio vulgaris* (common groundsel) Asteraceae
a cluster of young and mature heads; **b** half head; **c** young floret; **d** L.S. ovary of older floret; **e** T.S. ovary, older floret, placentation basal; **f** fruit; **g** young floret—upper part of corolla opened to show stamens and style-arms; **h** stamens (a–f ×6, g–h ×12)

ASTERACEAE

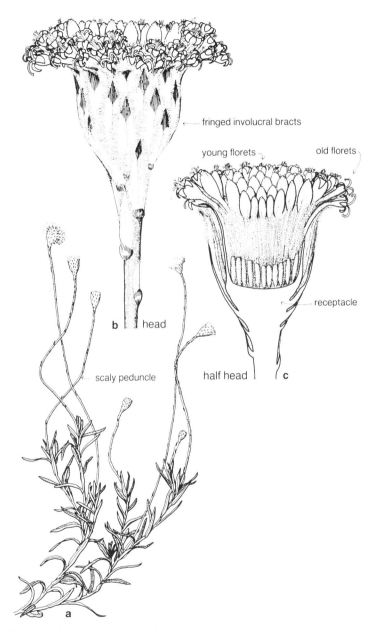

Fig. 107 *Leptorhynchos squamatus* (scaly buttons) Asteraceae (a ×0.5, b–c ×6)

Plant Families

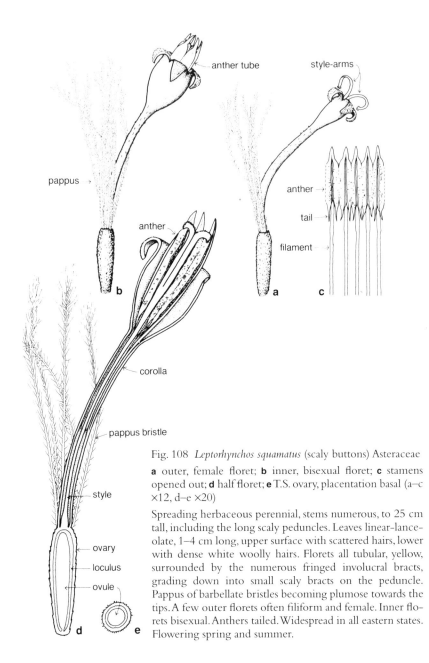

Fig. 108 *Leptorhynchos squamatus* (scaly buttons) Asteraceae
a outer, female floret; **b** inner, bisexual floret; **c** stamens opened out; **d** half floret; **e** T.S. ovary, placentation basal (a–c ×12, d–e ×20)

Spreading herbaceous perennial, stems numerous, to 25 cm tall, including the long scaly peduncles. Leaves linear-lanceolate, 1–4 cm long, upper surface with scattered hairs, lower with dense white woolly hairs. Florets all tubular, yellow, surrounded by the numerous fringed involucral bracts, grading down into small scaly bracts on the peduncle. Pappus of barbellate bristles becoming plumose towards the tips. A few outer florets often filiform and female. Inner florets bisexual. Anthers tailed. Widespread in all eastern states. Flowering spring and summer.

ASTERACEAE

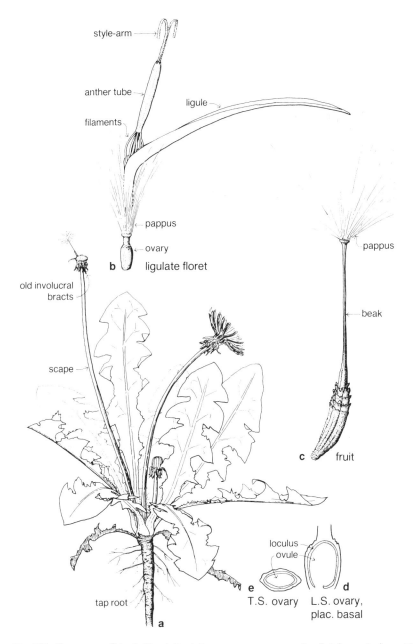

Fig. 109 *Taraxacum officinale* (dandelion) Asteraceae (a ×0.6, b–c ×6, d–e ×12)

205

Plant Families

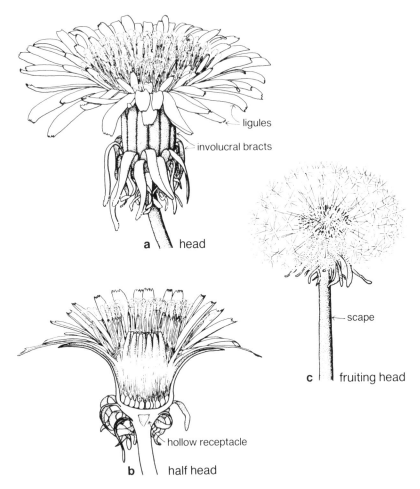

Fig. 110 *Taraxacum officinale* (dandelion) Asteraceae

Herbaceous perennial, with a well-developed tap root. Leaves in a basal rosette, pinnatifid and often described as runcinate, 3–15 cm long with pointed lobes. Flower heads with yellow ligulate florets only, borne on single unbranched scapes that exude latex when broken. Involucral bracts numerous, herbaceous. Pappus of numerous capillary bristles. Cypselas (sometimes called achenes) beaked. Widespread naturalised weed native to Europe and Asia, now virtually cosmopolitan. Recent research has made some headway in distinguishing various dandelions naturalised in Victoria (see bibliography no. 219, vol. 4, p. 688) all of which have previously been included under *T. officinale* in what is known as a species aggregate. Flowering mostly spring and summer. (a–c ×3)

ASTERACEAE

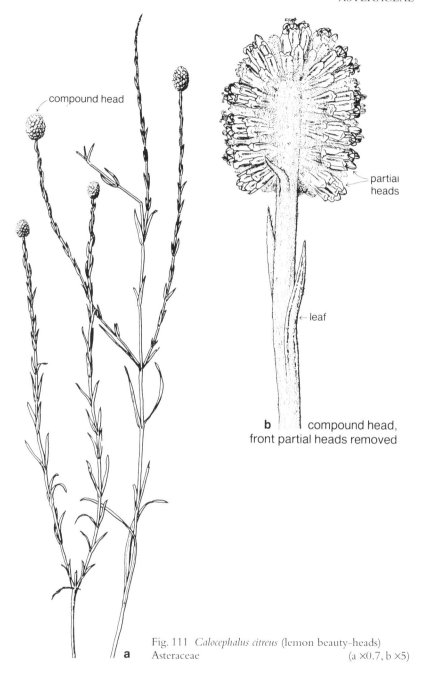

Fig. 111 *Calocephalus citreus* (lemon beauty-heads)
Asteraceae (a ×0.7, b ×5)

Plant Families

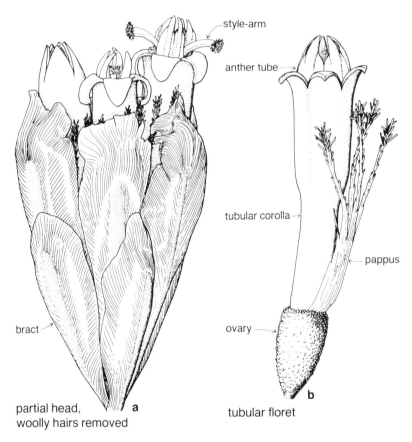

Fig. 112 *Calocephalus citreus* (lemon beauty-heads) Asteraceae
Perennial herb to 40 cm high. Stems branching, silvery grey and densely hairy. Leaves mostly opposite, linear, to 8 cm long, covered with fine white hairs. Compound flower heads yellow. Partial heads small with few tubular florets, immersed in woolly hairs. Pappus of fine scales with plumose tips, united at the base. Cypsela (sometimes called an achene) papillose. Widespread in lowland areas of all states except WA. Flowering late spring and summer. (a–b ×25)

Name that Flower

Plates

Plate 1 Floral structure: ovary superior, petals free

a *Crassula multicava* (crassula) Crassulaceae   K4 C4 A4 G$\underline{4}$

Top: flower, from above, showing the usual alternating parts, i.e. stamens alternating with petals, and carpels with stamens (cf. Pl. 1b). The sepals are small and hidden from view. Bottom: flower, side view, with the front parts removed. Flowers are < 1cm across. The Crassulaceae are widely cultivated; this species is a low, scrambling, herbaceous perennial native to South Africa. Flowering in spring (cf. Fig. 13).

b *Hibbertia empetrifolia* (guinea-flower) Dilleniaceae   K5 C5 A∞ G$\underline{2}$

Top: flower, from above. Stamens are grouped on one side of the 2 free carpels. Notched yellow petals are typical of the genus. Bottom: flower, side view, with the front perianth parts removed; the hairy ovaries are visible. One ovule can be seen in the loculus of the right hand ovary where the wall has been cut open. Small straggling shrub with flowers ± 1 cm across. Widespread in eastern Australia, flowering in spring. Sometimes listed under the name *H. astrotricha* (cf. Fig. 75).

c *Boronia denticulata* (boronia) Rutaceae   K4 C4 A4+4 G($\underline{4}$)

Top: flower, from above, showing the deep pink disc surrounding the green gynoecium. Bottom: side view of a flower which has been cut vertically and about half removed. Two loculi are visible; the ovules have been removed from the right hand one. Stamens in 2 whorls, filaments of 2 lengths, anthers dorsifixed. Small shrub native to WA and often cultivated; flowers ± 1 cm across appearing in spring (cf. Fig. 62).

d *Tetratheca ciliata* (pink-bells) Tremandraceae   K4 C4 A8 G($\underline{2}$)

Flower, side view, with 2 petals, 4 stamens and half the ovary removed. Anthers basifixed, dehiscing by terminal pores. Carpels 2, completely united, style 1. Ovary densely hairy, loculi 2, the ovules have been removed from the right hand one. See also Fig. 40.

e *Hypericum leschenaultii* (hypericum, St. John's wort) Hypericaceae   K5 C5 A∞ G($\underline{5}$)

Detail of flower—one yellow petal, and one group of stamens have been removed. The pale-green gynoecium, ± 1.8 cm long, is prominent in the centre of the flower. The style has 5 branches, each tipped by a small stigma. Stamens numerous, grouped into 5 bundles. A green sepal is visible at the base of the flower. Slender shrub to 3 m tall with opposite leaves. A native of Indonesia, this and other species are common in cultivation. Flowering over a long period. The genus is sometimes included in the family Guttiferae.

f *Ranunculus* sp. (buttercup) Ranunculaceae   K5 C5 A∞ G∞

Flower, right, and young fruit, left. Buttercup flowers have many stamens and numerous small, green, free carpels—more easily distinguished in the fruiting stage. The flower develops into an aggregate fruit; each fruitlet is called an achene. See also the notes on the family, p. 92.

Plate 2 Floral structure: ovary superior, petals / tepals united

a *Grevillea rosmarinifolia* (rosemary grevillea) Proteaceae P($\widehat{4}$) A4 G$\underline{1}$
Flower, side view, with half the perianth removed. The green ovary has been cut to show the loculus from which the ovules have been removed. A white nectary gland is present at the base of the ovary. Anthers ± sessile, epitepalous, showing emerging pale pollen. See also Fig. 33.

b *Grevillea* sp. (grevillea) Proteaceae P($\widehat{4}$) A4 G$\underline{1}$
A raceme of flowers, opening in sequence from buds with styles emerging, right, to open flowers. Perianth covered with appressed silky hairs (cf. Figs. 31–3).

c *Physalis viscosa* (sticky ground-cherry) Solanaceae K(5) C($\widehat{5}$) A5 G($\underline{2}$)
Tangled stems with flowers and fruit. Herbaceous rhizomic perennial, about 30 cm tall. Lower leaves ovate-lanceolate, about 5 × 3 cm, upper leaves lanceolate, smaller. Flowers solitary in leaf axils. Sepals united, calyx persisting in the fruiting stage, becoming strongly ribbed and eventually colouring orange-yellow and completely enclosing the berry. Corolla funnel-shaped, here about 2 cm diameter, with very small pointed lobes. Native to North and South America, a naturalized weed in south-eastern Australia. Flowering in summer and autumn. See also the notes on the family, p. 183, and Plate 8.

d *Epacris impressa* (common heath) Epacridaceae K5 C($\widehat{5}$) A5 G($\underline{5}$)
Part of flowering shoot. Flowers pendulous, solitary in leaf axils. Corolla tubular, with 5 small indentations ('impressions') near the base. The 5 free sepals grade into numerous smaller bracts on the pedicels. The purple anthers and the green stigma are visible in the throat of the corolla of some flowers. See also Figs. 88–9.

e *Acrotriche serrulata* (honey-pots) Epacridaceae K5 C($\widehat{5}$) A5 G($\underline{5}$)
Part of flowering shoot. Flowers borne in short congested spikes on the older stems, often concealed by younger growth. Sepals green, with purplish tips. Corolla tubular, pale-green, with a tuft of hairs near the apex of each lobe. Old pale-brown anthers are visible between the spreading lobes of some corollas. The flowers produce copious nectar, a characteristic suggested by the common name. Low spreading shrub forming mats up to 75 cm across. Leaves lanceolate, about 1 cm long, parallel veins visible on the lower surface. Widespread in woodlands in south-eastern Australia, flowering in spring. See also the section on the family, p. 171.

f *Nicotiana alata* (nicotiana, wild tobacco) Solanaceae
T.S. flower. The flowers are similar in general features to *Physalis* (Pl. 2c). The section has cut the outer, green, glandular hairy calyx, the pale-brown corolla, and the green ovary. Within the ovary, the horizontal septum divides the two loculi, each occupied by numerous small ovules arranged in a horseshoe shape on the domed placenta. Two of the ovules have been dislodged and lie on the ovary wall (bottom right). A yellow nectary disc surrounds the base of the ovary. See also the notes on the family, p. 183, and Plate 8.

g *Rhododendron laetum* (rhododendron) Ericaceae
Spray of flowers. The small rim-like calyx is hidden at the base of the flower. Corolla narrowly funnel-shaped with spreading lobes, up to 7 cm long and 5 cm wide. Stamens usually 10, the light-brown anthers dehiscing by pores, the pollen cohering in white 'threads'. The green style and stigma are visible in the centre of the flower. Slender leggy shrub to about 2 m tall, sometimes cultivated but hybrids are more commonly grown. Native to New Guinea. Flowering time variable in cultivation, mainly autumn. See also p. 171.

Plate 3 Floral structure: **a** ovary superior, petals free; **b–d** ovary inferior

a *Papaver nudicaule* (Iceland poppy) Papaveraceae K2 C4 A∞ G($\underline{6}$)
Tufted herbaceous perennial. Leaves radical, ± glaucous, pinnately lobed, up to 10 cm long. Flowers up to 7 cm diameter, borne on erect, slender scapes. Stiff hairs present on scapes and deciduous sepals. The yellow stigmas are sessile and in the shape of a star on top of the ovary, i.e. there is no discernible style. Native to the arctic regions of the northern hemisphere. Double forms and a variety of colours are grown.

b *Leucojum aestivum* (snowflake) Amaryllidaceae P($\overparen{3+3}$) A6 G($\overline{3}$)
Dissected flower, with the green tip of the spathe (right), and the pedicel of another flower. The style has a green band near the tip. The yellow anthers are basifixed. Three white ovules are visible in the dissected loculus. See also Fig. 124.

c *Scaevola pallida* (coast fan-flower) Goodeniaceae K5 C(5) A5 G(loculus $\overline{1}$)
Two flowers, side view. Flowers about 1 cm long, zygomorphic, subtended by a leaf-like bract and two bracteoles. The bract and one bracteole have been removed from the right hand flower. The calyx is a small, green, lobed rim at the base of the white and yellow corolla which is split down the back with the lobes spreading like a fan. Indusium present at the top of the style. Small, procumbent shrub, common along the coast in south-eastern Australia, flowering spring to early summer (cf. Fig. 99).

d *Leptospermum scoparium* 'Lambethii' (manuka) Myrtaceae K5 C5 A∞ G(-5-) floral tube present
Flower, from above, and dissected flower, side view. Flowers about 1.5 cm across, actinomorphic. Bases of perianth parts and stamens arising from the rim of the floral tube. Petals orbicular, pale to deep pink. Ovary semi-inferior to inferior, the top domed and covered by dark red nectary tissue which also lines the floral tube. Dense shrub to 3 m tall, mostly with white flowers in wild forms. The species is native to NSW, Vic., and Tas., as well as New Zealand where most of the pink-flowered cultivars have originated. Common in cultivation, double forms often grown (cf. Fig. 86).

Plate 4 Plants with unusual flowers or form

a–b *Amyema pendulum* (drooping mistletoe) Loranthaceae
a Flowering stems. The ovaries of some flowers are unusually swollen and may have been infected by a gall-producing insect. **b** Mature fruit, top right, and germinating seeds. At the end of the elongating green 'arm' the sucker-like haustorium will form on contact with the branch. Widespread stem parasite in south-eastern Australia, found on many native and introduced host species. See also the notes on the family, p. 87.

c–d *Allocasuarina verticillata* (drooping she-oak) Casuarinaceae
c Flowering branch of female plant. Fruiting cones are about 4 cm long. Inset: head of female flowers of *Allocasuarina* sp. with the dull reddish styles exserted. **d** Top: flowering branch showing elongated spikes of male flowers. Bottom: part of two flower spikes. The yellow anthers are about 3 mm long. Dioecious tree, 5–9 m tall, often with a drooping habit. Widespread in coastal south-eastern Australia, also the western basalt plains in Vic. Flowering in winter. Previously known as *Casuarina stricta*. See also the notes on the family, p. 72.

e *Drosera glanduligera* (scarlet sundew) Droseraceae
Small herb with glandular hairy leaves in a basal rosette, each leaf about 1.5 cm long with a narrow petiole and $\pm$ orbicular blade. Flowering stems hairy, up to 6 cm tall. Corolla orange-red. Widespread in Vic. on sandy soils, also in WA, SA, and NSW. Flowering in late winter to spring. See also the notes on the family, p. 95.

f *Cassytha melantha* (coarse dodder-laurel) Lauraceae
Twining parasitic perennial, forming haustoria where the stems contact those of a potential host. Flowers borne in short spikes, developing into a fleshy fruit about 1 cm in diameter (immature in this picture). The outer part of the fruit is the enlarged floral tube and is crowned by the dark remains of the perianth. Common in drier forests of WA, SA, Vic., NSW and Tas. Flowering in winter to spring. See also the notes on the family, p. 94.

Plate 5 Family CHENOPODIACEAE (saltbushes and samphires)

a *Atriplex cinerea* (coast saltbush)

Leafy flowering branches of a female plant (left) and male plant (right). The female flower (inset, lower) consists only of the ovary and two reddish stigmas. It is enclosed within two opposite ovate bracteoles which enlarge as the fruit matures (inset, upper—front bracteole removed). The male flowers are aggregated into dense clusters; their structure is similar to that shown in Pl. 6d. Leaves ovate to oblong, up to 8 cm long, 2.5 cm wide. Upright or spreading greyish shrub to about 1.5 m tall, widespread along the coast of southern Australia from WA to NSW and Tas. Flowering mainly late spring to summer.

The genus *Atriplex*, of more than 250 species, is distributed world-wide, with about 60 species in Australia known generally as saltbushes. The fruiting bracteoles vary considerably in size, shape, and ornamentation, and provide many features used in identification of species.

b *Suaeda australis* (austral seablite) P(5) A5 G(2)

Leafy flowering stems. Leaves are succulent and ± terete, mostly 7–20 mm long, with small clusters of flowers in the upper axils; mature buds are about 2 mm in diameter. Flowers usually bisexual, the five yellow anthers exserted from the perianth when mature (inset, left), followed by the stigmas. Plants vary in colour—green, yellowish or reddish. Bushy perennial to 70 cm tall, locally common in coastal and estuarine saltmarsh, all states except NT. Flowering mostly summer to autumn.

c *Maireana triptera* (three-wing bluebush)

Flowering and fruiting branch. The fruiting perianth has a short tube (not visible here). Attached to the apex of the tube is a circular 'horizontal' papery glabrous wing to about 10 mm diameter, pinkish-green at first, dark brown to blackish when mature. A further 3–5 'vertical' wings extend down the tube. Dark, thread-like remains of the slender stigmas project from the centres of the wings in some of the fruits. Leaves fleshy, subterete, 7–20 mm long. Compact shrub usually to about 50 cm tall. Widespread in inland Australia, in all mainland states. Flowering mainly winter to early summer.

d *Halosarcia pluriflora* (samphire)

Fruiting branches. The small flowers are embedded in the fleshy axis forming a terminal spike (much like *Sarcocornia*, Fig. 37) which dries out when mature. Three to seven flowers are borne above each of the opposite bracts. Several small dark seeds can be seen in the disintegrating spike. Small shrub to about 1 m tall with succulent, jointed, leafless branches, mostly occurring in central and northern SA, around salt lakes and springs, also in northern NSW. Flowering at any time of year.

e *Enchylaena tomentosa* (ruby saltbush)

Leafy fruiting stems. Fruiting perianth succulent, depressed globular, about 5 mm diameter, and coloured greenish to yellow or red (yellow in this picture), drying out to be black and shrivelled. Cf. *Rhagodia* (Pl. 6e) in which the ovary becomes succulent. Leaves terete, up to 20 mm long, glaucous or hairy. Lax shrub to about 1 m, or semi-prostrate, widespread, occurring in all mainland states. Flowering and fruiting at any time of year. The genus, of 2 species, is endemic to Australia.

f *Maireana sedifolia* (bluebush, pearl bluebush)

Plants in inland South Australia. Shrubs are dioecious, and grow to about 1 m high. Long-lived and drought resistant with deep tap roots, they provide useful fodder in dry periods when more palatable grasses and herbs are scarce. Occurs in all mainland states. The genus, of about 60 species, is endemic to Australia, the members generally known as bluebushes.

g *Tecticornia verrucosa* (tecticornia)

Plants growing in a dry claypan at Rainbow Valley in NT. Waxy greyish-green leafless stems are succulent and about 6 mm diameter. Flowering branches are borne at right angles to the main stems (inset), the flowers occurring usually in groups of three concealed between the succulent overlapping bracts. Stamens and small whitish stigmas are exserted as the flowers mature. Annual or short-lived perennial to about 40 cm tall, with scattered distribution in WA, SA and NT. Flowering time variable.

See also the notes on the family, p. 88, and Pl. 6d, e.

Plate 6 Various families

a *Euphorbia wulfenii* (spurge) Euphorbiaceae

Inflorescences. Each inflorescence of tiny flowers is surrounded by bracts which bear 4 crescent-shaped glands that are yellow when mature. The style can be seen protruding between the glands in some of the inflorescences. Each style has 3 branches, and each stigma is also branched. Pale-green orbicular bracts encircle the stems below the inflorescences. The broken edge of one bract is exuding droplets of white latex. Dense shrub to about 1.2 m high, with bluish-green linear to oblanceolate leaves, up to 10 cm or more long. Native to the Mediterranean region, commonly cultivated (cf. Fig. 69). Sometimes listed as *E. characias* ssp. *wulfenii*.

b *Zantedeschia aethiopica* (arum lily, calla) Araceae

Inflorescence and spathe. The orange inflorescence, called a spadix, is a fleshy spike 5–10 cm long, bearing male flowers in the upper part and female below. The male flower is represented by a few stamens only and the female flower by a small green gynoecium, some of which are seen protruding from the spike. Perianth parts are absent. The white bract, or spathe, is up to 25 cm long. Tufted, herbaceous perennial with sagittate leaves, native to South Africa. Widely cultivated, naturalised in WA and SE Australia. Flowering mainly in spring and summer.

c *Brunonia australis* (blue pincushion) Brunoniaceae

Flowering plants. The inflorescence is a head that superficially resembles a daisy. Flowering time is late spring and summer. See also the notes on the family, p. 191.

d *Rhagodia candolleana* (seaberry saltbush) Chenopodiaceae   P5 A5 G0

Part of a male inflorescence. At the top and centre right are open flowers with exserted stamens, at the lower left, one from which the anthers have fallen. Small, green non-functional ovaries can be seen in the lower and right hand flowers. The tepals are covered with whitish globular hairs that may collapse forming a mealy or scurfy covering. Weak scrambling shrub to 5 m tall, with small flowers, ± 3 mm across, in terminal panicles. Common in coastal situations in south-eastern Australia. Flowering mainly summer to autumn. Previously known as *R. baccata*. See also the notes on the family, p. 88.

e *Rhagodia parabolica* (fragrant saltbush) Chenopodiaceae   P5 A0 G($\overline{2}$)

Part of a female inflorescence, with fruit. The succulent berry is about 4 mm across, dark red when mature, and contains a single dark seed. Small flowers with withered pale-brown stigmas can also be seen. Soft shrub to 2 m tall mainly on rocky hillsides and creek banks, with a rather scattered distribution in central, southern and eastern Australia. Flowering mainly in winter. See also the notes on the family, p. 88.

f *Melaleuca armillaris* (giant honey-myrtle) Myrtaceae   K5 C5 A(∞) G($\overline{3}$) floral tube present

Inflorescence and fruit. Flowers sessile, forming a spike about 6 cm long. Stamens numerous, united into 5 bundles in each flower, the bundles alternating with the small white petals. The pale-green style and stigma can be seen emerging from the buds at the top of the spike. Fruit a woody capsule, persisting on the plant for some years. Bushy shrub or small tree 2–14 m tall, with alternate, narrow-linear leaves to 3 cm long. Originally confined to coastal parts of eastern Australia but now widely cultivated, flowering mainly spring to summer (cf. Fig. 87).

Plate 7 Legumes

a *Acacia genistifolia* (spreading wattle) Mimosaceae
Part of flowering stem. Flowers borne in axillary globular heads on slender peduncles. The numerous yellow stamens completely hide the other floral parts (cf. Pl. 7b). Erect spreading shrub, with slender, rigid phyllodes. Widespread in heathlands and drier forests in Vic., NSW, and Tas. Flowering in early spring. Previously known as *A. diffusa*.

b *Acacia myrtifolia* (myrtle wattle) Mimosaceae
Part of flowering shoot. Two open flowers of one head, and two buds of another, are visible in the short axillary raceme. The buds show the rim-like calyx surrounding the 4 unopened petals. In the open flowers, the petals spread out below the numerous stamens. Phyllodes lanceolate-oblanceolate, with a prominent midvein. The upper phyllode has two small stipules at the base (these will later fall), and a gland on the upper margin. Erect shrub to 2 m high, common on coastal and near-coastal heathlands from WA to Qld, and Tas. Flowering late winter to spring.

c *Acacia longifolia* var. *sophorae* (coast wattle) Mimosaceae
Flowering branches. Bushy spreading shrub, 1–3 m high. Phyllodes up to 10×3.5 cm, with several parallel veins. Flowers arranged in axillary spikes. Common along the coast from SA to Qld, also in Tas. Flowering late winter to spring. Some authors treat this variety at specific rank as *A. sophorae*.

d *Trifolium repens* (white clover) Fabaceae
Inflorescence. Not obviously a member of the pea family, but the individual flowers show the typical structure. The green calyx has 5 narrow, pointed lobes. The inflorescence is often loosely called a head even though the flowers are stalked. Herbaceous perennial with creeping stems which root at the nodes. Originated in Europe, now extensively cultivated as a pasture or lawn plant, often escaping as a weed. Occurs in all states except NT. Flowering mainly spring and summer.

e *Dillwynia sericea* (showy parrot-pea) Fabaceae
Part of flowering branch. Erect, heath-like shrub to 1.5 m high. Leaves terete, to 2 cm long, often rough and hairy. Flowers about 1 cm across, subsessile in upper leaf axils, forming leafy racemes. Calyx hairy. The petal at the back of the flower is broad compared with those of *Pultenaea* or *Daviesia*, for example, which often have flowers of a similar colour. Widespread in heathlands of SA, NSW, Vic. and Tas. Flowering in spring (cf. Fig. 55).

See also the sections on the legume families beginning on p. 109.

Plate 8 Family SOLANACEAE (nightshades)

The flowers are often distinctive with sepals united, petals united, stamens epipetalous, and anthers, although not united, often arranged in a neat ring around the style. Floral formulae are typically K(5) C($\widehat{5}$) A5 G($\underline{2}$); in *Brugmansia* the calyx is spathe-like and splits longitudinally.

a *Cestrum parqui* (green poison-berry)
Leafy flowering stems, and young fruit. Inflorescence a panicle. Calyx tubular, short. Corolla greenish-yellow, about 16 mm long, the limb with 5 $\pm$ triangular lobes. Leaves 3.5–5 cm long, paler green below. Fruits maturing glossy black. Woody shrub to about 3 m tall, suckering, toxic to animals. Native to S America, cultivated in gardens, now a naturalised weed. Flowering spring and summer.

b *Solanum nigrum* (black nightshade)
Stem with fruits and flowers. Inflorescences racemose with 4–8 flowers. Corolla white, stellate, about 1 cm across; anthers erect. Style slender, exserted between the anthers, stigma green, capitate. Fruits maturing purplish black. Leaves ovate, 4–7 cm long. Plants to about 0.5 m tall, often dark purplish-green, mainly herbaceous but may live for several seasons. European, but now a common weed. Flowering all seasons.

c *Solanum laciniatum* (large kangaroo-apple)
Left: flower, buds, leaves. Centre: flower with part of calyx, corolla and two stamens removed to show the ovary and style, and stamens attached to corolla tube. Right: Berry (1.5–2 cm diam.) and leaves. Corolla bluish-purple, about 3.5 cm across. Leaves usually lanceolate if entire, or broader and variously lobed. Short-lived shrub to 3 m tall, found in SA, Vic., and Tas. Flowering in summer.

d *Solanum ellipticum* (velvet potato bush, potato weed)
Leafy branches with flowers and fruit. Corolla rotate, purple, about 20 mm diameter. Berry becoming yellowish-green, calyx enlarging. Leaves ovate to elliptic, densely stellate hairy, mostly 4–8 cm long. Variable, sprawling perennial herb or subshrub to 1 m diameter, often with prickles on stems, petioles and peduncles. Widespread in central Australia. Flowering at any time but mostly early spring.

e *Lycopersicon* 'Burnley Sure Crop' (tomato)
Part of an inflorescence with flowers and young fruit. Corolla yellow, about 25 mm across, anthers prominent in the centre. Calyx persistent on the green berry (lower right) which will become a typical red, $\pm$ globular tomato. Branched, aromatic annual, widely cultivated (one of many cultivars), originally from S America. Flowering in summer.

f *Salpichroa origanifolia* (pampas lily of the valley)
Leafy stems, white flowers (about 7 mm long) and immature fruit. Inset: flower with front of corolla removed to show the dark red disc below the small conical greenish ovary, and white, hairy style. Leaves $\pm$ ovate to about 2 cm long. Rampant scrambling perennial with a dense vigorous root system. Origin S America, a weed in all states except NT. Flowering spring to autumn.

g *Brugmansia* x *candida* (angel's trumpet)
Leaves and pendulous flowers. Calyx spathe-like, splitting on one side. Corolla to 30 cm long, with 5 pointed lobes. Shrub or small tree of hybrid origin, to about 4 m tall, with ovate-oblong-elliptic leaves to about 20 cm long. Flowering in summer and autumn. Previously in the genus *Datura*.

h *Duboisia hopwoodii* (pituri)
Leafy flowering branches. Corollas pale, here about 6 mm across (often larger) striped internally with purple. Some flowers show the small bilobed stigma in the centre dusted with pale pollen. Bushy shrub to about 4 m tall, with narrow leaves up to 12 cm long, 1.3 cm wide. Widespread but scattered in arid areas of WA, southern NT and SA, extending west to NSW and Qld. Flowering mostly in spring.

i *Nicotiana suaveolens* (austral tobacco)
Flowering plant. Corolla white, pubescent, glandular, fragrant, up to 4.5 cm or more long; stamens and style included. Erect herb to 1.5 m tall, in Vic. and NSW. Flowers mainly spring and summer.
See also the section on the family, p. 183.

Plate 9 Various families including MYOPORACEAE (myoporums and emu-bushes)

In *Eremophila* and *Myoporum*, floral formulae take the form K(5) or K5 C$\widehat{(5)\ A4}$ G($\underline{2}$) but number of loculi varies. See also the section on the family, p. 188.

a *Myoporum* sp. (sticky boobialla) Myoporaceae
Leafy flowering branches. Flowers 2–7 in axillary clusters with hairy white corollas, here about 12 mm across, spotted with purple. Stamens strongly exserted, with purplish anthers. The glossy green fruits are immature, clasped by the narrow calyx lobes. Leaves ovate to lanceolate, 2–9 cm long, 0.6–4 cm wide. Erect shrub to about 2 m tall, occuring in SA, and Vic. This species, as yet without a formal name, was previously included under *M. viscosum*. Flowering mostly in late winter to spring.

b *Myoporum insulare* (common boobialla) Myoporaceae
Left: Leafy flowering branch. Right: Flower from above (top), and flower with one sepal and front part of corolla removed (below). Flowers 3–8 in axillary clusters, with hairy white corollas here about 8 mm across. In the dissected flower the central green conical ovary supports a pale whitish style and stigma; one calyx lobe (about 1.5 mm long) to the right of the ovary. Members of this family usually have four stamens, but some flowers on this plant have five. Leaves lanceolate to broadly elliptic, mostly 3–9 cm long. Bushy shrub or small tree in coastal vegetation from WA west to NSW and Tas. Often planted as a windbreak. Flowering mostly spring and early summer.

c *Eremophila glabra* (common emu-bush, tar bush) Myoporaceae
Leafy flowering branches. The red corolla of the flower lower left has detached to reveal the slightly expanded pale yellowish base of the corolla tube which earlier surrounded the ovary within the calyx. Calyx lobes similar in colour to the leaves. Corollas have one lobe strongly recurved. Anthers brownish, long exserted on pale filaments. Glossy green immature fruits have the slender styles still attached. A very variable species. Plants can be prostrate, to upright shrubs to 3 m tall, and flowers can be green, yellow, orange, red or brown. Leaves linear to (ob)lanceolate, about 1–6 cm long, 0.1–1.8 cm wide. Occurs in inland areas of mainland states. This form is indigenous to inland WA, and known in horticulture by the cultivar name 'Murchison Magic'. Flowering mainly in spring.

d *Eremophila maculata* (spotted emu-bush) Myoporaceae
Leafy flowering branch and one old fruit. Flowers solitary in the axils, the pedicels usually distinctively curved. Corollas have one lobe strongly recurved. Anthers brownish, strongly exserted on pale filaments. Fruit finally dry and woody, up to 18 mm long. Leaves linear to oblanceolate, mostly 1–4 cm long, 2–4 mm wide. Rounded much-branched shrub to about 2 m tall. Occurs in all mainland states. Flowering mostly late autumn to early summer.

e *Eremophila ovata* (emu-bush) Myoporaceae
Leafy flowering branches. Corollas mauve-blue, upper lip of 2 lobes, lower lip with 3. Stamens included. Leaves usually ovate, 1–2.5 cm long, densely stellate hairy. Rounded shrub to about 1 m tall, restricted to two ranges in southern NT. Flowering as conditions allow.

f *Eremophila platycalyx* (emu-bush) Myoporaceae
Leafy flowering branches. Calyces pink and showy. Corollas tubular, slightly curved, pale pinkish, with the upper lip of 2 lobes, the lower lip with 3. Bushy shrub to about 2 m tall, occurring in arid inland areas of WA, SA and NT. Flowering mainly in spring.

g *Hibiscus rosa-sinensis* (rose-of-China, Chinese hibiscus) Malvaceae   K(5) C5 $\widehat{A(\infty)}$ G($\underline{5}$)
Single flower. The 5 united sepals are surrounded by linear epicalyx lobes. Petals pink, joined to the base of the filament column which rises in the centre of the flower and carries yellow anthers on the free ends of the filaments. The column encloses the ovary and style, the latter protruding at the top, and dividing into five branches, each with a globular stigma. Robust shrub originally from tropical Asia, flowering mostly spring and summer. A wide range of cultivars is commonly grown, many of hybrid origin; the botanical name is often applied rather loosely. (Cf. Fig. 11.)

h *Stylidium graminifolium* (grass triggerplant) Stylidiaceae   K(5) C(5) A$\overline{2}$ $\overline{G}$(loculi $\overline{1}$ or $\overline{2}$)
Part of an inflorescence with several flowers. Greenish calyx lobes are united into two lips. Dark-tipped glandular hairs dot the stem and calyces. Corolla tube with erect white appendages at the throat. Corolla lobes bright pink, four larger and spreading, one (the labellum) smaller and usually recurved. Staminal filaments united with the style to form a column that is irritable (i.e., responds to touch), here bent back in the 'set' position. Found in SA, Vic., and Tas. Flowering in spring. See the notes on the family p. 191, and Fig. 100 which illustrates a similar species.

Plate 10 Family ASTERACEAE (daisies) **a–e** radiate heads, **f–h** ligulate heads, **i** compound head

a *Galinsoga parviflora* (gallant soldier)
Radiate heads: left, side view; centre, cut longitudinally; right, top view. Five small, white ray florets can be seen at the edge of the head. Central disc florets yellow, numerous. Each floret is subtended by a small scale, $\pm$ 2.5 mm long, seen against the dark ovaries in the centre head. The pappus is of fringed scales. Small, widespread, annual weed with heads $\pm$ 5 mm across, and opposite leaves. Introduced from South America. Flowering most of the year.

b *Chrysanthemoides monilifera* (boneseed)
Bushy shrub to 2 m tall. Young leaves and stems with woolly hairs. Ray florets 4–8, fertile, producing the rather fleshy, dark green to purple fruit. Disc florets numerous, functionally male. Native to South Africa, now a serious weed in all states except NT. Flowering spring to summer.

c *Senecio elegans* (purple groundsel)
Erect annual about 30 cm tall, with pinnatifid leaves. Involucral bracts in one main row, with extra dark-tipped bracts at the base of the head. Ligulate florets purple, sometimes white, disc florets yellow. Pappus of numerous fine bristles: the head is a fluffy ball in the fruiting stage, with the withered yellow-brown corollas of the disc florets still visible. Native to South Africa, and now naturalised in coastal southern Australia. Flowering in summer (cf. Figs. 105–6).

d *Calendula officinalis* (garden marigold)
Radiate head, from above, showing the typical structure with the youngest florets in the centre, progressively older to the edge of the head. Annual or short-lived perennial herb to 50 cm tall, native to southern Europe. Commonly grown and often escaping from cultivation. Many cultivars and double forms are known. Flowering most of the year.

e *Podolepis jaceoides* (showy podolepis)
Radiate head, from above. The ligules are deeply 3-lobed. Youngest florets are in the centre of the head. Perennial herb to 70 cm tall, widespread in Vic., SA, NSW, Qld and Tas. Flowering in spring and early summer.

f–g *Sonchus oleraceus* (sow thistle, milk thistle)
f Ligulate heads: left, external view; right, cut longitudinally, exposing the florets. **g** Ligulate florets. The small rounded ovary is at the base, below the pappus of fine hairs which surround the white, hairy corolla tube. The tube extends above into the flattened yellow ligule. Widespread annual weed to 1 m tall, heads $\pm$ 1 cm long. Probably native to Europe, flowering most of the year.

h *Hypochoeris radicata* (flatweed)
Ligulate head, from above. The plant is very similar to a dandelion (Fig. 109) with leaves in a basal rosette, but the flowering stems are usually branched. Native to Europe and now a widespread weed, often in lawns, flowering most of the year.

i *Pycnosorus chrysanthes* (golden billy-buttons)
Flowering stems with compound heads. Annual or perennial herb to about 40 cm tall, scattered on flat open country in Vic., SA and NSW, flowering mainly spring and summer. Previously known as *Craspedia chrysantha*. (Cf. Figs. 111–12.)

j *Cynara scolymus* (artichoke)
Herbaceous perennial up to 2 m tall, cultivated for the fleshy receptacle of the flower head and the thickened bases of the involucral bracts, which are edible. The lower leaves are deeply dissected. Native to the Mediterranean.

See also the section on the family, p. 193.

Plate 11 Monocotyledons, Part 1

a–c *Avena sterilis* (sterile oat) Poaceae

a Small inflorescence (left), and an old inflorescence from which the fruits have fallen, only the bleached glumes remaining. Tufted annual or short-lived perennial to about 1 m tall, inflorescence an open panicle. **b** Spikelet, showing the two green glumes enclosing two florets, of which only the pale, hairy lemmas are visible. On the back of each lemma is a stout dark awn. **c** Young floret, with feathery white stigma and part of a pale-yellow anther protruding between the lemma and palea. The slender stalk at the right is the rachilla which ends in a rudimentary floret. Native to Europe, now a widespread weed, flowering mainly spring and summer. See also the section on the family, p. 211.

d *Alopecurus pratensis* (meadow fox-tail) Poaceae

Inflorescences. Tufted perennial to about 1 m tall. Inflorescence a dense, spike-like panicle about 6 cm long. The lemma bears a slender dorsal awn. At different stages, the feathery stigmas (left) or stamens are conspicuous. Anthers are about 3 mm long and carried on slender filaments. Native to Europe and northern Asia, an occasional weed in Australia. Flowering in spring and summer. See also the section on the family, p. 211.

e *Thysanotus tuberosus* (common fringe-lily) Liliaceae   P3+3 A6 G($\underline{3}$)

Flowers from above. The typical lily structure is disguised by the fringed inner tepals, and curved style and stamens. Inflorescence paniculate, to 60 cm tall. Tufted perennial with linear leaves, found in SA, Qld, NSW, and Vic. Flowering in summer. See also the section on the family, p. 219.

f *Wolffia australiana* (wolffia) Lemnaceae

Top: Three plants, side view. Bottom: plants from above. This tiny floating aquatic maintains the green (photosynthetic) side uppermost. Plant body (called a thallus) here about 1 mm long; roots and leaves absent. Reproduction commonly vegetative; each plant 'buds off' a daughter—in this way colonies form rapidly on still water bodies. Flowers are extremely reduced—the male to a single stamen with a pale yellowish unilocular anther, the female to a single unilocular ovary within the plant body, the rounded stigma exserted beside the anther. The species is not uncommon in SA, NSW, Vic., Tas., and NZ. The genus includes the smallest known flowering plants.

g *Caleana minor* (small duck-orchid) Orchidaceae

Flower, side view. Most conspicuous is the glossy dark purplish, warty labellum about 6 mm long held above the rest of the flower on its narrow, curved, green claw. The other perianth parts are narrow and project down to the left. The column, also projecting down to the left, is held ± at right angles to the inferior ovary, has broad green lateral wings, and ends in the yellowish anther. Plants are glabrous, up to about 15 cm tall with a reddish-brown linear leaf 4–13 cm long. Occurs in all states except WA, flowering in spring and summer. See also the section on the family p. 230.

h *Prasophyllum spicatum* (leek-orchid) Orchidaceae

Part of an inflorescence, with several flowers shown from the front and the side. Compared to most orchids in southern Australia, these flowers appear upside down. The lateral petals are 6–8 mm long, and white with a central purplish stripe. At the top of the flower is the conspicuous white labellum, bent sharply upwards at the middle, with the margin intensely folded and jagged. The purple-striped dorsal sepal is at the bottom of the flower. Flowering stem slender, to 80 cm tall, with flower spike up to 20 cm long. A single cylindrical leaf up to 50 cm long partly sheaths the base of the flowering stem. Occurs in southern Vic. (west of Wilsons Promontory) and SA. Flowering in spring, most freely after fire or other disturbance. See also the section on the family p. 230.

i *Thelymitra aristata* (sun orchid) Orchidaceae

Part of an inflorescence. Sepals and petals similar, purple-blue, strongly veined (labellum not differentiated). Column with 3 prominent lobes, the two side lobes each with a brush of white hairs, and the rear lobe curved forward and tipped yellow. The anther is hidden behind the rounded whitish stigma visible within the base of the column. The small pointed projection at the top of the stigma is the rostellum. Robust herb 35–75 cm tall with a single, broad-lanceolate leaf. Stem-bracts 3 or 4 below the inflorescence, a raceme of numerous flowers, each about 3 cm across. Widespread, although not common, in Qld, NSW, SA, Vic., and Tas. In Willis (bibliography no. 233, vol. 1), this species is listed under the name *T. grandiflora*. See also the section on the family, p. 230.

Plate 12 Monocotyledons, Part 2

a *Lepidosperma gladiatum* (coast sword-sedge) and *Ficinia nodosa* (knobby club-rush) Cyperaceae
Flowering plants. The sword-sedge is behind the club-rush. See also captions **d** and **g**.

b–c *Juncus sarophorus* (broom rush) Juncaceae
b Flowering plant. Leaves are reduced to basal sheaths in this species, straw coloured above grading to dark brown below. The colour of basal sheaths is an important feature in identification of 'leafless' *Juncus* species. **c** Part of stem and inflorescence. Flowers, although small brownish and often densely clustered, are separate in this family, not aggregated into spikelets as in the sedges or grasses. Each flower has six straw-coloured, pointed tepals seen here clasping the developing fruit (a capsule). The stems are longitudinally striated (the number of striations varies with species), and each stem appears to continue beyond its inflorescence; this projection is called the primary bract. Densely tufted perennial, occurring in SA, Vic., NSW, Tas., NZ. Growing in consistently damp sites, flowering mostly spring and summer.

d *Ficinia nodosa* (knobby club-rush) Cyperaceae
Main picture: inflorescences. Inset: inflorescence, side view, with front spikelets removed. Bisexual flowers (florets) with no perianth are each subtended by one bract (called a glume) and spirally arranged in spikelets, the spikelets densely clustered in a globose or hemispherical head about 1–2 cm diameter at the apex of the culm. The culm appears to continue beyond the inflorescence and end in a sharp point; this continuation is referred to as a bract. Numerous pale yellowish anthers are shown exserted from the reddish-brown glumes and form the most conspicuous parts of the flowers. Strongly rhizomic perennial with tufts of stems (culms) up to 1 m tall arising at intervals (Pl. 12a). 'Leaves' are reduced to brownish sheaths at the bases of the culms. A locally common species particularly of near-coastal sandy soils or fringing saline water bodies, or in sandy heathlands, in all states except NT. Flowering in spring. Previously known as *Isolepis nodosa*.

e–f *Cyperus eragrostis* (drain flat-sedge, umbrella sedge) Cyperaceae
e Flowering stems. Spikelets about 3 mm wide, clustered at the ends of the inflorescence branches. Several long, leaf-like bracts arise at the base of the inflorescence. **f** Several spikelets, side view. Green bracts (all called glumes), each about 2.5 mm long, are arranged regularly in two rows, one on each side of the central spikelet axis. Each glume encloses a flower which consists of a stamen and a gynoecium; the perianth is absent. One yellow anther can be seen exserted from a glume at the top left, and developing pale brownish-yellow fruits can be seen within the lower glumes, with one fruit dislodged and pointing down to the left. Tufted grass-like perennial usually 30–50 cm tall, with stems triangular in T.S., and linear leaves. Native to America, a widespread weed in southern and eastern Australia. Flowering mainly in summer to autumn. See also p. 213.

g *Lepidosperma gladiatum* (coast sword-sedge) Cyperaceae
Main picture: old (left) and young inflorescences. Inset: Several spikelets on a branch of an inflorescence. The inflorescences are made up of numerous spikelets about 8 mm long, each of which contains several tough brown bracts (all called glumes) spirally arranged around the spikelet axis. One of the upper glumes subtends a bisexual flower, another lower down a male flower, and the other glumes are empty. The bisexual flower is made up of 3 stamens and a gynoecium. A perianth is absent but is represented by 6 small whitish scales below the ovary. In the inset lower left the withered, thread-like remains of three style arms are folded back over the glumes. A mature brownish-yellow fruit (a nut) protrudes from the upper floret; the pale base of the style persists on the top. Tufted, clump-forming perennial (Pl. 12a), with tough upright stems and leaves, widespread along the coast of southern Australia from WA to NSW and Tas., mostly growing on sand dunes. Flowering spring–summer.

h–i *Xanthorrhoea australis* (austral grass-tree) Xanthorrhoeaceae
h Flowering plants. Trunk stout, up to 3 m high, bearing a dense tuft of linear leaves, with a terminal spike-like inflorescence to 2.5 m long. **i** Part of inflorescence. Flowers creamy white, surrounded by brown bracts which obscure the perianths. Stamens 6, prominently exserted. Widespread in dry forests and heathlands in Vic., SA, NSW and Tas. Flowering winter to summer but usually only after fire. See also the section on the family, p. 218.

Monocotyledons

Most monocots are herbaceous annuals or perennials that shoot each season from an underground bulb, corm or tuberous rhizome, and many are aquatic or marsh plants. Some, such as the palms and grass-trees, are arborescent. Many species have very short stems, and most leaves are basal, sometimes forming dense tussocks. These leaves are usually long and slender and have parallel venation. The leaves of erect species are often sessile with stem-clasping bases, and may have parallel or reticulate venation.

The flower parts are frequently in threes, although in some families this is modified by reduction. When the perianth is petaloid there are usually two whorls, each of three parts, which are similar in shape, size and texture, called **tepals**. The terms sepals and petals are seldom used except in the orchids. The perianth is much reduced or absent in many families such as the grasses, sedges and rushes.

Many changes have occurred in the delimitation of monocot families and genera in recent revisions, and the situation remains unresolved. The family Amaryllidaceae has been combined with Liliaceae (and this has been adopted for the new *Flora of Australia*), and the genus *Allium* (onions and garlics) is either assigned to one or other of these families or placed in a separate one, Alliaceae. *Xanthorrhoea* (grass-trees) and *Lomandra* (mat-rushes) are here placed in Xanthorrhoeaceae, but will be found under Liliaceae in many books, and there are numerous other examples. This book treats Liliaceae and Amaryllidaceae separately, and gives a short account of Iridaceae (irises), Orchidaceae (orchids) and Xanthorrhoeaceae. It contains a brief introduction to the grasses and for those wishing further information some references are included.

POACEAE grasses

(GRAMINEAE)

Although the grass family is not as large as the daisy family in terms of numbers of genera and species, grasses are found almost everywhere (even Antarctica), and are estimated to comprise about 20 per cent of the world's vegetation cover. The family name is derived from the large genus *Poa*, which includes over 40 native species, many of which are conspicuous tussock grasses of forests and grasslands. The genus also includes weeds and lawn grasses.

Economically, Poaceae is the most important plant family, not only as a source of food for humans, but also for grazing animals. Many species, such as wheat, barley, maize and rice, have been subjected to breeding and selection in order to improve yield and disease resistance. In a number of countries, bamboo stems and leaves, and the leaves of other large grasses, are used as building materials.

Aboriginal people used many native grasses as a food source. They usually ground the seeds to a flour, which they then mixed with water and baked.

There is considerable variation in form within the family, and numerous classification schemes have been proposed. A current scheme recognises six sub-families, further divided into 40 tribes.

Members of the family are illustrated in Figures 113–17 and Plate 11a–d. For keys to species of Australian grasses, see bibliography number 206 and CD5. References for grasses in various regions include, for central Australia, bibliography number 144; New South Wales, number 222; the Northern Territory, number 207; Queensland, number 214; Tasmania, number 165; Victoria, number 71. Other works include numbers 118 and 141.

FLORAL STRUCTURE

Flowers   Small, usually bisexual, sometimes unisexual or sterile, and together with 2 enclosing bracts are known as **florets** (Fig. 117; Pl. 11c).

Perianth   Either considered absent or represented by the lodicules, which are small colourless scales, usually 2, at the base of the ovary.

Androecium   Stamens usually 3, rarely 1–6 or more, almost always free.

Gynoecium   Ovary with 1 loculus and 1 more or less basal ovule. Carpel number a matter of debate. Styles and stigmas usually 2.

Fruit   Usually a caryopsis.

Grasses are perennial or annual herbs, sometimes tree-like (bamboos), sometimes forming tussocks, or with rhizomes or stolons. The root system is fibrous, and large grasses such as bamboo or maize often produce adventitious roots, or prop roots, from the lower stem nodes. The upper stems of grasses, usually called **culms**, terminate in an inflorescence.

Plant Families

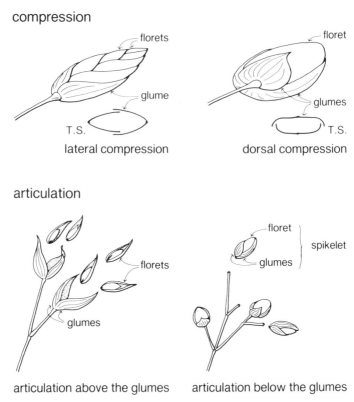

Fig. 113 Grasses—compression and articulation
Grass spikelets usually lie readily on their sides or backs: lateral or dorsal compression. At maturity, the glumes either fall with the spikelet, or they remain attached to the plant: articulation below or above the glumes.

The leaf usually has a basal **sheath**, almost always open down one side, which is wrapped around the stem. The leaf **blade** is linear, often flat, but may be rolled, and has parallel venation. At the junction of the sheath and blade is nearly always a line of hairs or a membranous flap of tissue called a **ligule** (Fig. 114b).

FURTHER NOTES ON THE FLOWERS AND INFLORESCENCES

The small flower together with its two subtending bracts is called a **floret** (Fig. 117; Pl. 11c). The outer or lower bract is the **lemma** or flowering glume, and this partially encloses the upper bract, the **palea**. Between the lemma and palea are the ovary and stamens. The tiny lodicules are at the base of the ovary next to the lemma, and at anthesis become swollen, pushing the lemma and palea apart and allowing the stamens and stigmas to protrude. The florets are commonly bisexual but may be sterile, male or (rarely) female. Sterile florets, also called barren or neuter florets, may be reduced to a lemma only. Rudimentary florets, where the bracts (lemma and palea) are reduced in size, sometimes to stubs of tissue, are not uncommon.

The florets are arranged in spikelets (Fig. 114d; Pl. 11b). The **spikelet** is made up of one or more sessile florets on an axis, the **rachilla**, and at the base are two further bracts, the upper and lower **glumes**. Sometimes one of these glumes is reduced in size; rarely one is absent.

The spikelet is considered as the basic unit of grass inflorescences. Like flowers of other families, spikelets may be pedicillate or sessile, and arranged in spikes, racemes, panicles or more complex inflorescences (Figs 114–16; Pl. 11a, d).

The term **compression** describes the way in which the spikelets are flattened (Fig. 113). It is lateral if the spikelet lies readily on its side, and dorsal if the spikelet is flattened back and front. Occasionally spikelets are terete.

At maturity the spikelet may break up, with the florets falling separately, or it may fall as a unit. When the glumes remain on the plant, the **articulation** is said to be 'above the glumes'. When they fall with the spikelet, articulation is 'below the glumes' (Fig. 113). This can often be seen before the fruiting stage by gently pulling the florets and noting the point of breakage.

Stiff, bristle-like appendages, called **awns**, are quite common, particularly on the lemmas (Pl. 11c, d).

The common terms 'grain' and 'husks' are not always used consistently. Typically the husks are the lemma and palea, and the grain is the fruit.

Superficially, grasses and sedges (in the family Cyperaceae, Pl. 12a, d, e, f, g) are quite similar. In the sedges, the leaf sheath is usually closed, and the ligule absent. The culms are often solid, and flattened or triangular in cross-section. The inflorescence is often subtended by obvious bracts. The florets are grouped in spikelets but each flower is enclosed by only one bract.

Summary of grass structure

'flower' + lemma + palea = floret
one or more florets + glumes (upper + lower) = spikelet
spikelet = the basic unit of grass inflorescences, arranged in spikes, racemes, panicles

SPOTTING CHARACTERS

Plants herbaceous, culms usually terete, with hollow internodes. Ligule usually present, leaf sheath open. Florets with two bracts (lemma and palea). Spikelet with two bracts (upper and lower glumes).

Plant Families

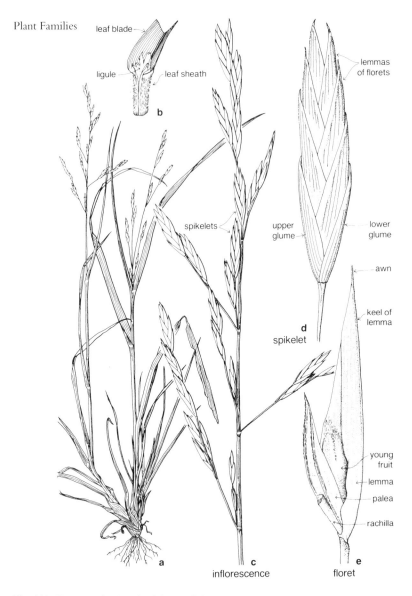

Fig. 114 *Bromus catharticus* (prairie grass) Poaceae
Annual or biennial, 40–100 cm high, stems stout and glabrous. Leaves rough on upper side, sheaths often softly hairy. Ligule membranous. Spikelets flat, 6- to 8-flowered, pale green, in a loose panicle. Compression lateral. Articulation above the glumes. Widespread in Australia, a common weed or pasture grass, introduced from South America. Flowering spring to early summer, or longer if conditions suitable. Often known by the name *Bromus unioloides*. (a ×0.2, b ×1.5, c ×0.6, d ×2.5, e ×4)

POACEAE

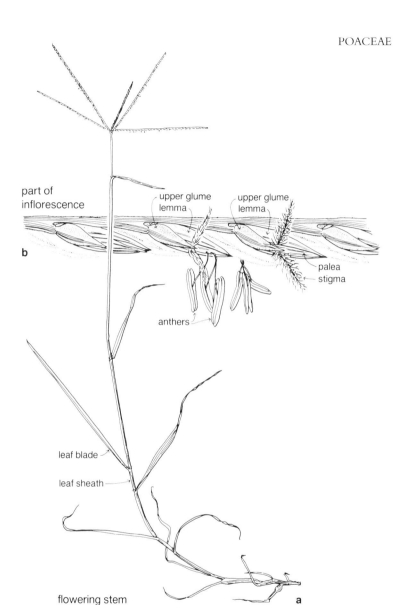

Fig. 115 *Cynodon dactylon* (couch grass) Poaceae
Stoloniferous perennial, often long creeping. Ligule a ring of hairs. Spikelets purplish-green, 1-flowered, sessile, in 2 rows along one side of an axis forming a slender spike; 2–6 of these spikes arranged digitately at the top of the culm. Compression lateral. Articulation above the glumes. Widespread in Australia, common as a weed or lawn grass. Flowering in summer. (a ×0.6, b ×12)

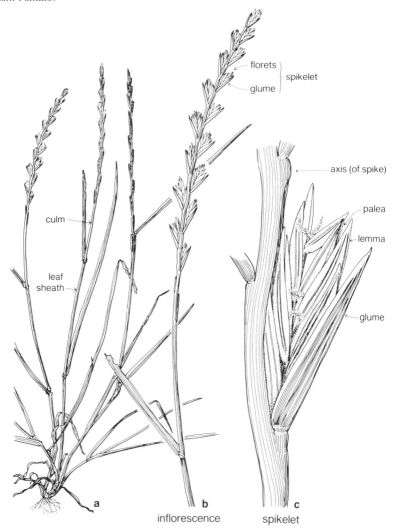

Fig. 116 *Lolium perenne* (perennial rye-grass) Poaceae
Biennial or short-lived perennial up to 50 cm high. Leaves narrow-linear, mainly basal. Leaf blades auriculate, ligule membranous. Spikelets about 8- to 10-flowered, solitary and sessile in alternate notches of the rachis, forming a slender spike. Topmost spikelet with two glumes, all others with only one. Compression lateral. Articulation above the glume. Native to Europe, North Africa and Asia, now naturalised in Australia, and a common and valuable pasture or lawn grass. Plants are often found with short awns on some of the lemmas. These are usually regarded as hybrids, many of which have been developed for agricultural purposes. Flowering in summer. (a ×0.5, b ×1, c ×6)

POACEAE

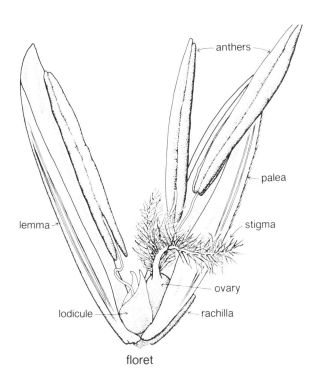

Fig. 117 *Lolium perenne* (perennial rye-grass) Poaceae (×12)

XANTHORRHOEACEAE grass-trees, mat-rushes

Of the ten genera and about 100 species in the family Xanthorrhoeaceae only two *Lomandra* spp. (mat-rush) and one *Romnalda* sp. extend outside Australia. Six genera are restricted to Western Australia, but the type genus *Xanthorrhoea* (grass-trees, Pl. 12h, i) is widespread in heathlands and drier forests. All members of the family are perennials, and most have more or less woody stems, some of them underground. The family is very diverse and some authors separate the genera into several families (e.g., *Flora of New South Wales*, bibliography no. 104).

For descriptions and keys to Australian species see *Flora of Australia* vol. 46 (bibliography no. 94).

FLORAL STRUCTURE

Flowers	Actinomorphic, bisexual, sometimes unisexual in *Lomandra*. Inflorescence variable: in *Xanthorrhoea* it is spike-like, while flowers are solitary in *Calectasia* (tinsel-lily).
Perianth	Tepals 6, in 2 whorls of 3, free or united.
Androecium	Stamens 6, sometimes epipetalous.
Gynoecium	Carpels 3, united. Ovary superior with 1–3 loculi. Style single, except for *Lomandra*, which has 3.
Fruit	Usually a capsule.

If a trunk is present it may be branched or single, with a rosette of leaves at the apex. When no trunk develops, the rosette is at ground level. The leaves are tough and linear, and in arborescent plants the older leaves may persist after they die, forming a brownish skirt underneath the terminal rosette. When the old leaves are finally shed, the resin-impregnated leaf bases remain attached to the trunk and these form a protective covering over the internal tissues. Flowering of some *Xanthorrhoea* spp. is stimulated by fire or defoliation.

Grass-trees were very useful to Aboriginal people, who ate the young shoots and soaked the inflorescences in water to yield a sweet drink. Resin from the leaf bases was used as an adhesive, and the stalks of the spikes had some value for spear shafts. In more recent times resin has been extracted from *Xanthorrhoea* spp. and used in the production of varnish and dyes.

LILIACEAE lily family

The cosmopolitan lily family is widespread, particularly in temperate regions. The family name is derived from the genus *Lilium*, which has one species naturalised in Australia although many have been introduced and are grown in gardens. Other introduced genera include *Tulipa* (tulip), *Lachenalia*, *Hyacinthus* (hyacinth) and *Muscari* (grape hyacinth). *Asparagus officinalis* (asparagus) is grown for its edible succulent shoots, but the scrambler *A. asparagoides*, introduced as an ornamental, has become a weed in many coastal districts. The tufted *Asphodelus fistulosus* (onion-weed) is also a locally common weed.

Native members of the family are common in the heaths and dryer forests of south-eastern Australia. These include *Chamaescilla* (blue stars, Fig. 119), *Arthropodium* (chocolate lily, Fig. 118), *Thelionema* (tufted blue-lily, Fig. 120), *Wurmbea* (early nancy, Fig. 121) and *Thysanotus* (fringe-lily, Pl. 11e).

For descriptions and keys to Australian species see *Flora of Australia* vol. 45 (bibliography no. 94).

FLORAL STRUCTURE

Flowers	Actinomorphic, usually bisexual. Inflorescence often a raceme, sometimes an umbel.
Perianth	Tepals 6, free or united, arranged in two whorls of 3.
Androecium	Stamens 6, sometimes epitepalous.
Gynoecium	Carpels 3, ovary superior, usually with 3 loculi. 1–many ovules per loculus. Placentation usually axile.
Fruit	Usually a capsule or berry.

The leaves, usually linear and entire, often arise in basal tufts, but on erect stems they are alternate. The rootstock may be a bulb, tuber, corm or rhizome.

SPOTTING CHARACTERS

Plants herbaceous. Leaves often linear and grass-like. Perianth petaloid, in 2 whorls of 3 parts. Stamens 6.

Plant Families

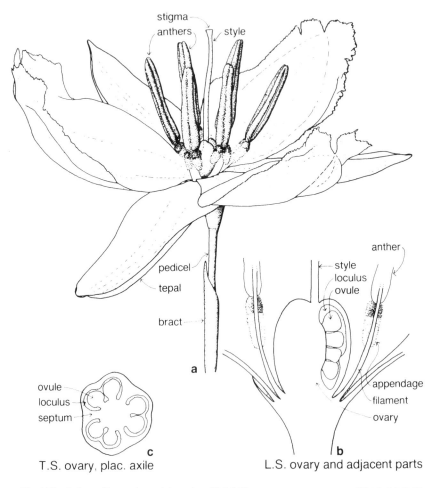

Fig. 118 *Arthropodium strictum* (chocolate lily) Liliaceae P3+3 A6 G($\underline{3}$)
Perennial herb to 1 m high. Leaves basal, linear, 7–30 cm long. Flowers blue or violet, scented, drooping, borne in racemes or loose panicles, one flower at each node. Anthers with prominent basal appendages. Fruit a capsule. Widespread in heaths, open forests and grasslands in all states. Flowering late spring to early summer. Sometimes known as *Dichopogon strictus*. (a ×7, b–c ×12)

LILIACEAE

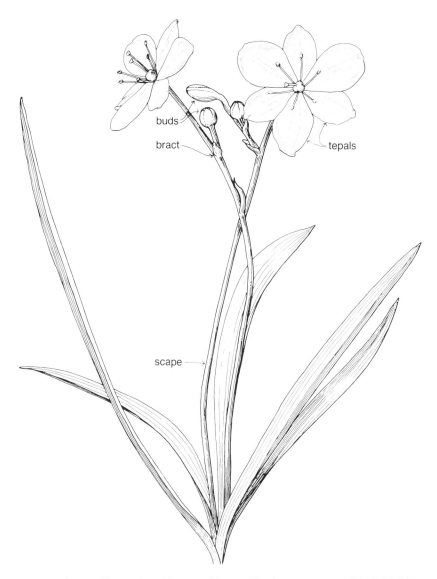

Fig. 119 *Chamaescilla corymbosa* (blue stars, blue squill) Liliaceae   P3+3 A6 G(3)
Perennial herb to 15 cm high. Leaves basal, linear, to 8 cm long. Flowers blue, borne in a loose corymb. Widespread and common on damp sandy ground, but not in the Alps. In all states except Qld and NT. Flowering in spring. (×1.5)

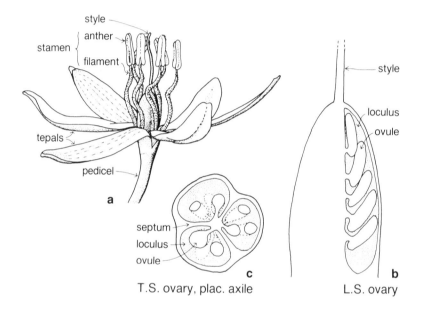

Fig. 120 *Thelionema caespitosum* (tufted blue-lily) Liliaceae   P3+3 A6 G(3)
Perennial herb to 60 cm high. Leaves in a basal tuft, narrow linear, to 25 cm long. Flowers blue, sometimes yellow or white, in open panicles. Filaments finely hairy. Stems leafless, with a few bracts. Fruit a capsule. Widespread in damp coastal heathlands and occasionally further inland. Eastern states. Flowering in spring. Previously known as *Stypandra caespitosa*. (a ×4, b–c ×12)

LILIACEAE

Name that Flower

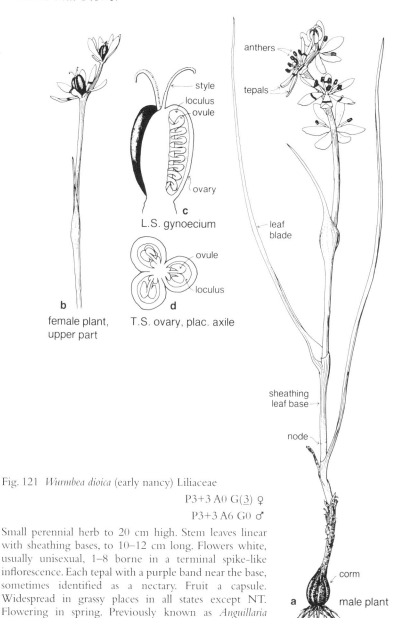

Fig. 121 *Wurmbea dioica* (early nancy) Liliaceae

P3+3 A0 G(3) ♀
P3+3 A6 G0 ♂

Small perennial herb to 20 cm high. Stem leaves linear with sheathing bases, to 10–12 cm long. Flowers white, usually unisexual, 1–8 borne in a terminal spike-like inflorescence. Each tepal with a purple band near the base, sometimes identified as a nectary. Fruit a capsule. Widespread in grassy places in all states except NT. Flowering in spring. Previously known as *Anguillaria dioica*. (a–b ×1.5, c–d ×10)

223

AMARYLLIDACEAE amaryllis family

The name of this large, cosmopolitan and horticulturally important family is derived from the South African genus *Amaryllis* (belladonna lily). The three genera native in Australia include *Crinum* (Murray lily) and *Calostemma* (garland lily). Many members of the family are sold as cut flowers, and numerous genera are common in cultivation. These include *Agapanthus* (Figs 122, 123), *Nerine*, *Amaryllis*, *Leucojum* (snowflake, Fig. 124) and *Narcissus* (daffodils and jonquils), and the last three are recorded as naturalised. The genus *Allium* includes cultivated onions, leeks and garlic as well as *A. vineale* (wild garlic) and *A. triquetrum* (three-cornered garlic), which are proclaimed weeds.

For descriptions and keys to native and naturalised species see *Flora of Australia* vol. 45 (bibliography no. 94) where they are included in the Liliaceae.

FLORAL STRUCTURE

Flowers	Actinomorphic, bisexual. Inflorescence usually an umbel, borne on a scape and subtended by one or more large, often papery bracts, each called a **spathe**.
Perianth	Tepals 6, free or united, arranged in 2 whorls of 3. The **corona** or trumpet of daffodils and jonquils is usually regarded as an outgrowth from the perianth.
Androecium	Stamens 6, sometimes epitepalous.
Gynoecium	Carpels 3. Ovary often inferior with 3 loculi. Ovules few to many per loculus. Placentation usually axile.
Fruit	A berry or capsule.

Most members of the family grow from a perennial bulb, which produces a cluster of basal linear leaves each season. In some species, such as *Amaryllis belladonna* (belladonna lily), the flowering stem appears before the leaves.

The family is closely related to the Liliaceae and is included in that family by some authors. The presence of an inferior ovary or, if the ovary is superior, of flowers in an umbel subtended by bracts is characteristic of the Amaryllidaceae.

SPOTTING CHARACTERS

Plants lily-like, with linear leaves arising from a bulb. Inflorescence often an umbel subtended by bracts.

AMARYLLIDACEAE

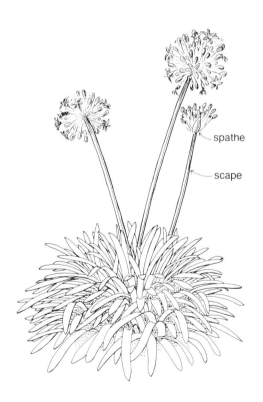

Fig. 122 *Agapanthus praecox* (agapanthus) Amaryllidaceae   (×0.05)

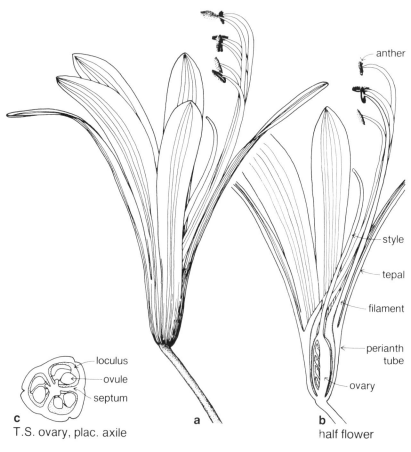

Fig. 123 *Agapanthus praecox* (agapanthus) Amaryllidaceae   P(3+3) A6 G($\underline{3}$)
Perennial, forming large clumps. Leaves radical, broad-linear up to 40 cm or more long. Flowers blue or white in a terminal umbel on a scape 60 cm or more tall. Tepals united at the base into a perianth tube. Fruit a capsule. Widely cultivated, sometimes weedy, origin South Africa. Flowering mainly in summer. (a–b ×2, c ×7)

AMARYLLIDACEAE

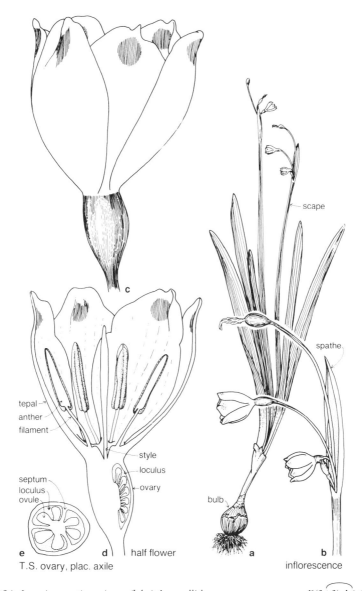

Fig. 124 *Leucojum aestivum* (snowflake) Amaryllidaceae P(3+3) A6 G($\overline{3}$)
Bulbous perennial. Leaves basal, linear about 20 cm long. Flowers nodding, borne in a small umbel. Tepals white, each with a green spot, often regarded as free, but united to a small degree at the base. Fruit a capsule. Widely cultivated, origin Europe. Flowering in early spring. See also Pl. 3b. (a ×0.25, b ×0.7, c–e ×4)

IRIDACEAE iris family

The family is found in all temperate and tropical parts of the world and is especially well represented in South Africa, and Central and South America. The type genus *Iris* is of European origin and is now naturalised in Australia. Many other introduced genera are familiar in gardens, including *Freesia, Gladiolus, Crocus, Sparaxis* and *Ixia*. Others, which were probably brought in for gardens, are now proclaimed weeds. The most widespread of these are *Homeria* (cape tulip), *Watsonia* and *Romulea longifolia* (onion grass). Native genera include *Patersonia* (purple-flag), *Libertia* (grass-flag, Fig. 125) and *Diplarrena* (butterfly flag or white iris).

For descriptions and keys to Australian species see *Flora of Australia* vol. 46 (bibliography no. 94).

FLORAL STRUCTURE

Flowers — Bisexual, either actinomorphic or zygomorphic. Inflorescences various, often paniculate. In *Freesia* and *Gladiolus* the inflorescence is spike-like and the individual flowers are sessile and subtended by 2 bracts.

Perianth — Tepals 6, often united into a tube as in *Freesia* and *Gladiolus*, free or almost so as in *Libertia* (Fig. 125), or the outer 3 united at the base to form a tube and the inner 3 erect in the centre, as in *Iris*.

Androecium — Stamens 3. Usually free, often unilateral, (i.e. filaments slightly curved so that the anthers lie on one side of the style).

Gynoecium — Carpels 3. Ovary inferior, usually with 3 loculi. Style single at the base, then dividing into 3 branches, which may be further divided. Ovules few to many. Placentation usually axile.

Fruit — A capsule.

Members of the family are mostly perennial herbs. The leaves form a rosette or tuft and arise from a bulb, corm or rhizome.

SPOTTING CHARACTERS

Plants herbaceous. Leaves often linear and grass-like. Perianth petaloid, in 2 whorls of 3 parts, sometimes zygomorphic. Stamens 3. Ovary inferior, of 3 united carpels.

IRIDACEAE

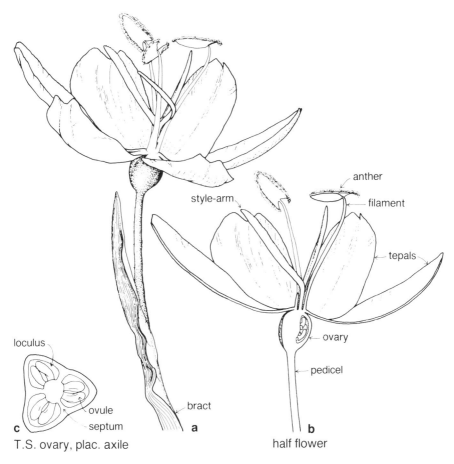

T.S. ovary, plac. axile   half flower

Fig. 125 *Libertia pulchella* (pretty grass-flag) Iridaceae   P3+3 A(3) G($\overline{3}$)
Perennial herb. Leaves mostly basal, narrow linear, to about 15 cm long. Flowering stems to 30 cm tall. Flowers white, on slender pedicels in a few loose clusters, forming a small panicle. Fruit a capsule. Found in damp shaded montane to subalpine forests of Vic., Tas. and NSW. Flowering in spring. (a–b ×7, c ×12)

ORCHIDACEAE orchid family

The Orchidaceae is a very large family of about 800 genera, of which about 100 are found in Australia. Although its members are spread all over the world, most of them are in the tropics.

The orchids are important commercially, as they are produced in considerable numbers for the cut-flower trade, and because orchid-growing is such a widespread hobby. Aboriginal people ate the root tubers of a large number of species, and in Arnhem Land orchid sap was used as a fixative for ochre pigments.

Numerous books have been produced on native orchids—see the list at the start of the bibliography. Comprehensive texts include numbers 12, 16, 70, 113 and 129. Number 22 is a useful field guide to species in south-eastern Australia. A classic work, now rather a collector's item, is Nicholls' *Orchids of Australia* (no. 167).

FLORAL STRUCTURE

Flowers   Zygomorphic (rarely more or less actinomorphic), bisexual. Flowers solitary, or inflorescence a raceme, spike or panicle (Figs 126–31; Pl. 11g–i).
Perianth   Parts 6, in 2 whorls, free, or united in various ways. The 3 segments of the outer whorl are called sepals. The 2 lateral segments of the inner whorl are called petals, and the remaining segment is the **labellum**.
Androecium   Stamens 1 or 2; 1 in almost all Australian species.
Gynoecium   The styles, stigmas and stamen(s) are united into a more or less erect fleshy structure, called the **column**, in the centre of the flower (Fig. 126). The ovary is inferior and unilocular, with 3 parietal placentas. There are numerous ovules (Fig. 127).
Fruit   A capsule.

Most orchids are perennial herbs. They are terrestrial, epiphytic or saprophytic and shoot annually from rhizomes, tubers or thickened rootstocks. Some form a pseudobulb, which is the thickened base of an aerial stem.

FURTHER NOTES ON THE FLOWER

The sepal that sits up at the back of the flower in most species is the dorsal or median sepal and it is often larger than the other two lateral ones. The modified third petal, the **labellum**, varies in shape, size and surface characters, and is usually at the front of the flower. It may be simple, toothed, lobed, or fringed with hairs, and the surface can be smooth, hairy (Fig. 127) or papillose, or with longer projections (Fig. 128b). Surface or marginal projections are often referred to as

ORCHIDACEAE

calli, and are sometimes glandular, e.g., secreting pheromones attractive to pollinators (Fig. 128b). A contraction of the tissue at the base of the labellum, if present, is called a claw (Fig. 126, Pl. 11g). In slipper orchids the labellum is shaped like a small sack.

There are one or two 2-lobed anthers attached near the top of the column. Nearly all Australian native orchids have 1 anther. The anther opens by longitudinal slits, and the pollen is often aggregated into groups called **pollinia** (Fig. 127). The end of a pollinium is sometimes drawn out to form a sterile stalk, the **caudicle**.

The stigma, which is borne on the part of the column facing the labellum, is often 2-lobed and there may be a small or conspicuous projection called the **rostellum**, between the stigma and the anther (Fig. 127). The column often has lateral outgrowths, which may be in the form of wings as in *Caladenia* (spider-orchid, Fig. 126) and *Pterostylis* (greenhood, Fig. 130), or hairy lobes as in *Thelymitra* (sun-orchid, Pl. 11i).

Some common orchids of southern Australia do not entirely fit in with the general pattern outlined above. For example, *Thelymitra* has a more or less actinomorphic flower with all the sepals and petals about the same size, and in *Diuris* the dorsal sepal is short and broad compared with the other two. In *Pterostylis* the dorsal sepal is more or less united to the two lateral petals to form a hood or galea (Figs 130, 131). In a few genera such as *Caleana* (duck-orchid, Pl. 11g), and *Prasophyllum* (midge-orchid, leek-orchid, Pl. 11h), the labellum is described as being above the column, which means that, compared with other orchids, the flower is upside-down. The column of *Diuris* is not a single structure, as the anther is borne on a short filament, and the stigma on a separate broad style. Above the stigma is a slot that is said to represent the rostellum. The slot encloses a viscid disc that, at maturity, adheres to the two pollinia from the anther.

SPOTTING CHARACTERS

Plants herbaceous. Flowers zygomorphic, with 1 petal, the labellum, different from the others, and a central column. Ovary inferior.

SELECTED EXAMPLES

Floral formulae have been omitted from the captions to Figures 126–31. They follow the general pattern K3 C3 A1 G($\overline{3}$), except for the greenhoods (Figs 130, 131), in which the perianth parts are variously united.

Plant Families

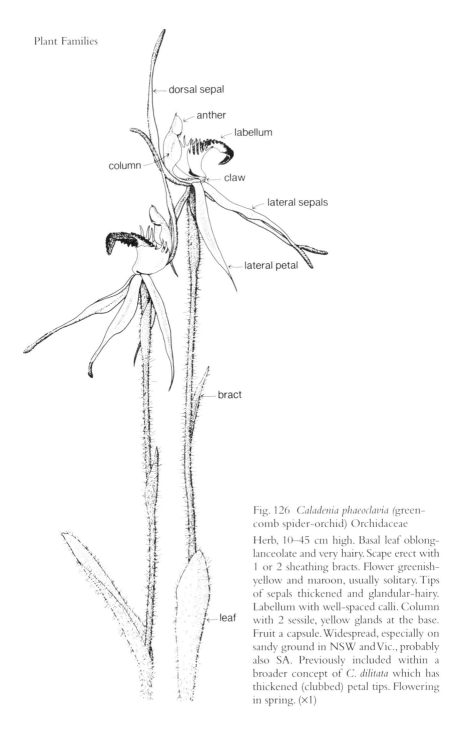

Fig. 126 *Caladenia phaeoclavia* (green-comb spider-orchid) Orchidaceae

Herb, 10–45 cm high. Basal leaf oblong-lanceolate and very hairy. Scape erect with 1 or 2 sheathing bracts. Flower greenish-yellow and maroon, usually solitary. Tips of sepals thickened and glandular-hairy. Labellum with well-spaced calli. Column with 2 sessile, yellow glands at the base. Fruit a capsule. Widespread, especially on sandy ground in NSW and Vic., probably also SA. Previously included within a broader concept of *C. dilitata* which has thickened (clubbed) petal tips. Flowering in spring. (×1)

ORCHIDACEAE

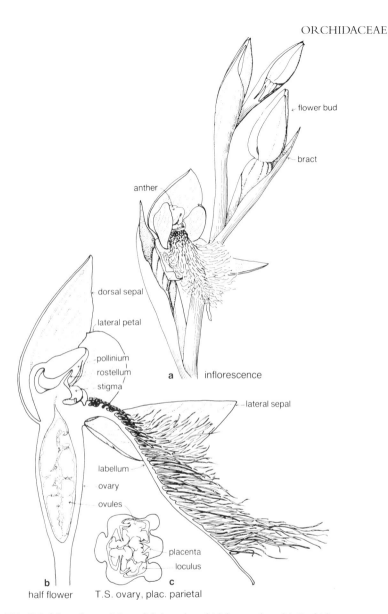

Fig. 127 *Calochilus robertsonii* (purplish beard-orchid, brown-beards) Orchidaceae
Herb, to about 40 cm high. Solitary basal leaf up to 20 cm long, broad at the base and tapering. Several leaf-like bracts on the scape, and one subtending each flower. Flowers green and purple, 2–9 in a loose raceme. Labellum densely bearded. The numerous ovules appear as a dense whitish mass. Fruit a capsule. Widespread in south-eastern Australia. Flowering in spring. (a ×2, b–c ×4)

233

Plant Families

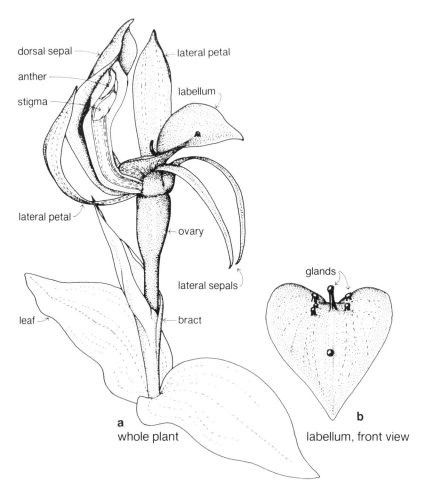

Fig. 128 *Chiloglottis valida* (common bird-orchid) Orchidaceae
Herb, short, about 5 cm high. Two basal leaves opposite, shortly-stalked, ovate, about 2 cm long. Scape with one sheathing bract. Flower solitary, bronzy-green or tinted with purplish-brown. Labellum with glandular calli (sometimes referred to as stalked and sessile glands). Widespread in highland forests of Vic. and NSW. Flowering in spring to summer. (a–b ×3)

ORCHIDACEAE

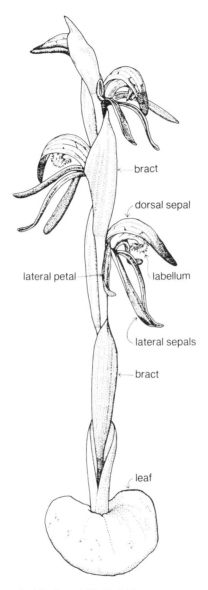

Fig. 129 *Pyrorchis nigricans* (red-beak orchid) Orchidaceae
Herb, 10–30 cm high. Leaf radical, broad, ovate to round, thick and fleshy, about 3–5 cm wide. Stem usually with two bracts. Flowers purple and sometimes streaked with purplish red, 2–8 in a racemose inflorescence. Flowers and stem turn black after fruiting. Widespread on damp and dry heaths. All states except Qld, NT. Flowering infrequent, stimulated by fire, usually spring. Previously known as *Lyperanthus nigricans*. (×1)

Plant Families

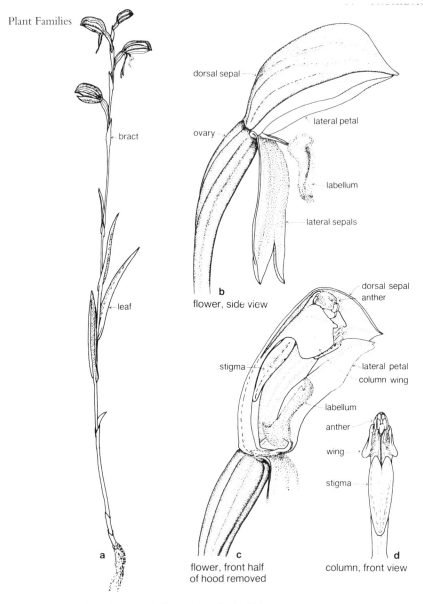

Fig. 130 *Pterostylis melagramma* (tall greenhood) Orchidaceae
Herb, 10–30 cm or more high. Stem leaves well developed, no basal rosette present at flowering time. Flowers green, 3–8 borne in a raceme. Lateral sepals bent downwards. Labellum irritable, flicking up into the hood when touched. Widespread in heathlands and forests in south-eastern Australia. Previously included with *P. longifolia*, a species now considered restricted to NSW. Flowering winter to spring. (a ×0.5, b–d ×4)

ORCHIDACEAE

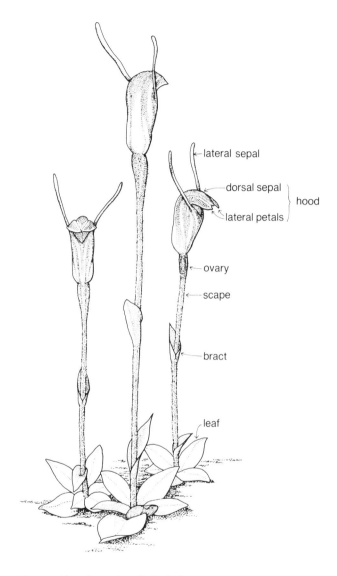

Fig. 131 *Pterostylis nana* (dwarf greenhood) Orchidaceae
Herb, usually less than 15 cm high. Leaves in a basal rosette, with short petioles, elliptical, 1–1.5 cm long. Flower green, solitary. Lateral sepals held erect. Widespread but never alpine, in NSW, Vic., and Tas. A complex of forms which in future may be split up. Flowering winter to spring. (×1)

Appendix

Nomenclatural Changes since the First Edition

The following name changes have come to our attention since *Name That Flower* was first published. The changes have been incorporated into the text of this edition. Names change for a number of reasons; it should not be assumed that the original names are all now synonyms.

Page number in first edn	Name used in first edn	Current Name
back cover	*Casuarina stricta*	*Allocasuarina verticillata*
vii plate 4	*Casuarina stricta*	*Allocasuarina verticillata*
plate 5	*Acacia sophorae*	*Acacia longifolia* var. *sophorae*
plate 7	*Craspedia chrysantha*	*Pycnosorus chrysanthes*
8 para 6	*Casuarina*	*Allocasuarina*
24 para 1	"	"
34 list	*Acacia sophorae*	*Acacia longifolia* var. *sophorae*
list	*Cassia*	*Senna*
list	*Craspedia*	*Pycnosorus*
38 list	*Cassia*	*Senna*
53 caption	*Casuarina stricta*	*Allocasuarina verticillata*
77 caption	*Hakea sericea*	*Hakea decurrens*
78 caption	"	"
90 para 2	*Pittosporum phillyreoides*	*Pittosporum angustifolium*
93 caption	"	"
95 para 2	*Acaena anserinifolia*	*Acaena novae-zelandiae*
101 para 1	*Albizia lophantha*	*Paraserianthes lophantha*
108 para 1	*Cassia*	*Senna*
para 4	" (within FLORAL STRUCTURE) "	

238

Appendix

Page number in first edn		Name used in first edn	Current Name
109	caption	*Cassia artemisioides*	*Senna artemisioides*
110	caption	"	"
120	para 2	*Microcitrus*	*Citrus*
	para 2	*Eremocitrus glauca*	*Citrus glauca*
	para 3	*Diosma*	*Coleonema*
124	caption	*Eriostemon myoporoides*	*Philotheca myoporoides*
136	caption	*Pomaderris oraria*	*Pomaderris paniculosa* ssp. *paralia*
139	caption	*Hibbertia stricta*	*Hibbertia riparia*
142	para 3	*Tristania*	*Lophostemon*
	lines 3 up	"	"
148	caption	*Acacia sophorae*	*Acacia longifolia* var. *sophorae*
170	para 2	*Lavandula denta*	*Lavandula dentata*
177	caption	*Stylidium graminifolium*	*Stylidium* sp. 2 (sens. *Flora of Vict.* 4: 587)
180	caption	*Craspedia chrysantha*	*Pycnosorus chrysanthes*
185	caption	*Helichrysum bracteatum*	*Xerochrysum bracteatum*
204	para 2	*Dichopogon*	*Arthropodium*
205	caption	*Dichopogon strictus*	*Arthropodium strictum*
207	caption	*Stypandra caespitosa*	*Thelionema caespitosum*
216	para 2	*Paracaleana*	*Caleana*
217	caption	*Caladenia dilatata*	*C. phaeoclavia*
219	caption	*Chiloglottis gunnii*	*C. valida*
220	caption	*Lyperanthus nigricans*	*Pyrorchis nigricans*
221	caption	*Pterostylis longifolia*	*Pterostylis melagramma*

Bibliography, CD-ROMs and Websites

This is not an exhaustive list, but includes most of the references useful for native and weedy flowering plant identification in Australia (with a bias towards the southern states) and several texts useful for New Zealand species. We have also included a small selection from the large number of books available that deal with cultivated plants.

CD-ROMs are beginning to have an impact on plant identification (see Chapter 7). To date, electronic keys have been produced for eucalypts and orchids of south-eastern Australia, Australian grasses and wattles, rainforest trees and shrubs, and families of flowering plants occurring in Australia. These are listed towards the end of the bibliography.

The internet has allowed unprecedented access to information generally, and many herbaria and botanic gardens have established websites. Users are now able to view much specialist literature electronically, for example a searchable version of the *Census of Victorian Vascular Plants* (bibliography no. 87) is available via the website of the Royal Botanic Gardens, Melbourne. A small selection of websites is given after the CD-ROMs at the end of the bibliography.

There is a considerable amount of information available in botanical journals. The CSIRO publishes the journal *Australian Systematic Botany*, and the herbaria in WA, SA, Qld, NSW, and Vic. each produce a journal, all of which include specialist papers dealing with various plant groups. The journals of the field naturalist clubs in each state often include useful articles of a more popular nature—for example, the series by Corrick in the *Victorian Naturalist* on the genus *Pultenaea* (bush-peas, no. 54). The journal of the Australian Plants Society (formerly the Society for Growing Australian Plants), called *Australian Plants*, is also helpful. Most journals provide an index at the end of a set of parts or volumes. Many journals are now available online. In the following list, about a dozen entries refer to journal articles and the rest are books.

Bibliography

To assist you to find a suitable text, various categories with appropriate reference numbers are listed below. Numbers in bold indicate current standard identification texts (mostly Floras or similar), or a few other books that we have found very useful in a particular category. Italic numbers indicate works (other than Floras) that include keys.

botany—standard textbooks 136, 143, 174, 181, 182
classification—general 94, 136, 143, 181, 182
Eucalyptus 35, 86, 111, 179
climbers 132, 133, 230
cultivated plants—exotic *13*, 14, 188–90, *209, 220*
native 78, *209*
dictionaries 15, 31, 96, **152**, 204
ecological
 alpine 51, *59*, 138, 155, 197
 aquatic *8*, 52, 61, 128, 186, 196, 211
 arid areas 64, **126**, 162, *213*, 217
 grasslands 74, 151, 154, 200
 mallee 158
 Melbourne sandbelt 201
 rainforest 53, *83–5, 89*, 105, 107, 130, 168, 230, 231, CD3
 Sydney sandstone 75
families and genera (see also *Flora of Australia*, no. **94**)
 Acacia (wattles) **94**, 103, *175*, 205, *212*, CD4, field guides 5, 185, *226*
 Allocasuarina (she-oaks) **94**
 Banksia 93, **94**, 115
 Brachyscome 198
 Caesalpiniaceae 68, **94**
 Callistemon (bottlebrushes) 240
 Cassia (*Senna*) **94**, 180
 Chenopodiaceae (saltbushes) **94**
 Conospermum (smoke-bushes) **94**, 195
 daisies 10
 Darwinia 26
 Drosera (sundews) 80, **94**, 149
 Dryandra **94**, 193
 Eremophila 208

Eucalyptus 36, 46, **57**, **94**, 116, 137, *213*, CD2
 classification 35, 86, 111, 179, field guides *37*, **58**, 146, *160*
 everlastings 10
 Fabaceae (pea family) 69, 101, 236, 238
 Goodeniaceae **94**
 Grevillea **94**, *159*, **169**, 239
 Hakea **94**, 239, 241
 Homoranthus 26
 Isopogon **94**, 194
 kangaroo paws 117
 Lambertia **94**, 195
 Lamiaceae (mint family) 124
 Leptospermum (tea-trees) 240
 Liliaceae (lily family) **94**
 mangroves *145*, 148, *202*
 Melaleuca 114, 240
 Orchidaceae (orchids) *12*, 16, 22, 47, 49, 70, 88, 97, 113, 129, 134, *135*, 167, 177, 192, 237, CD1
 palms 131
 Petrophile **94**, 194
 Poaceae (grasses) *71*, *118*, 141, *144*, *165*, *206*–7, *214*, *222*, CD5
 Prostanthera (mint-bushes) 2
 Proteaceae 27, 92, **94**, 194–5, 239
 Pultenaea (bush-peas) 54
 Restionaceae (scale-rushes, rope-rushes etc.) *161*, 203
 rushes and sedges (Qld) 203
 Rutaceae 6, 7
 Senna (*Cassia*) **94**, 180
 Stylidium (trigger-plants) 81
 Verticordia 95
 Westringia 3

241

Bibliography

families—general 13, 32, 57, 63, **109**,
110, *140*, **143**, 164, 182, 221, *CD6*
Floras (and a few other standard
identification manuals)
ACT **40**
Australia **94**
central Australia **126**
Kosciusko **59**
NSW 17, 18, **42**, **104**
NT **61**, **73**
New Zealand **1**
offshore islands **94**
Qld (south-east) **210**
SA 23, 24, 76, **127**
Tas. **65**
Vic. **219**, 233
WA **25**, **100**, **153**, **224–5**
fruits 53, 105
garden plants (see cultivated plants)
glossaries 67, 147, 218, W3 (many Floras
include glossaries)
historical 43, 44
insectivorous plants 80, *149*
lists of species
Australia 112
NSW 20, 123, W10
NT 45, 60, 72
New Zealand 171
Qld 108
SA 125
Tas. 39
Vic. 87
WA 98, 170, W5
nomenclature 31, 96, 99, 103, 150, 215
picture books 34, 41, 50, 55, 56, 79,
155–8, 187, 232, 234–5
plant structure 33, 143
regional (see also Floras—excluded from
this list)
Australia
south-east **57**
central 217
NSW *4*, 19, 21, 51, **57**, 62, 64, 74–5,
82, *83–5*, *89*, 90–1, 155, 168, *184*,
187, 230–1, 235
NT 34, 213, 217, 228
New Zealand 128, 191, 199, 221
Qld *48*, 53, 101, 107, 139, 187, *203*,
211, 231–2
SA *30*, **57**, 66, 91, 178
Tas. 41, 50, 120–1, 138
Vic. 11, *32*, 50, 56, **57–8**, 77, 91, 106,
155–8, 200, 216, 227, *233*, 234,
CD7
WA 38, 55, 79, 92, 119, 162, 170, 176,
183, 223
trees *4*, 29, *30*, **37**, **57–8**, *83–5*, *89*, 90,
102, 116, 137, 199, 213
weeds 9, 119–21, 139, 142, 163, 191,
197, 229
environmental 28, 166
noxious 121, 172–3
wildflowers 10, 38, 41, 50–1, 55–6, 66,
79, 91, 106, 155–8, 178, 187, 216–17,
223, 232, 234–5

Books and Articles

1 Allan, H. H., *Flora of New Zealand*. (Govt Printer, Wellington, NZ, 1961.) Keys and descriptions for all then-known indigenous vascular plants except monocots. Vol 2 (1970), edited by Moore, L. B., and Edgar, E., deals with indigenous monocots except grasses. Vol. 3 (1980), edited by Healy, A. J., and Edgar, E., deals with adventive monocots in 23 families (excludes grasses and some others). Vol. 4 (Botany Division, D.S.I.R., Christchurch, 1988), edited by Webb, C. J. *et al.*, deals with naturalised vascular plants except monocots. Vol. 5 (Manaaki Whenua Press, Lincoln, 2000) edited by Edgar, E., and Connor, H. E., deals with grasses.

Bibliography

2 Althofer, G. W., *Cradle of Incense. The Story of Australian* Prostanthera. (Soc. for Growing Aust. Plants, 1978.)
3 Althofer, G. W., and Althofer, P. McD., 'The Genus *Westringia*'. *Australian Plants*, 10: 343–59 (1980).
4 Anderson, R. H., *The Trees of New South Wales*. (Govt Printer, Sydney, 4th edn, 1968.) Includes keys and descriptions.
5 Armitage, I., *Acacias of New South Wales*. (Soc. for Growing Aust. Plants, 1977.) A useful field guide.
6 Armstrong, J. A., 'The Family Rutaceae', *Australian Plants*, 8: 195–9, 212–14 (1975).
7 Armstrong, J. A., 'The Family Rutaceae in Australia', *Australian Plants*, 8: 215–25 (1975).
8 Aston, H. I., *Aquatic Plants of Australia*. (Melbourne Uni. Press, Carlton 1977.) Comprehensive, well-illustrated manual with keys.
9 Auld, B. A., and Medd, R. W., *Weeds. An Illustrated Botanical Guide to the Weeds of Australia*. (Inkata Press, Melbourne, 1987.) A comprehensive work with many colour photographs. Species are arranged in families.
10 Australian Daisy Study Group, *Everlasting Daisies of Australia*. (C. H. Jerram and Assoc., with R. G. & F. J. Richardson, Meredith, 2002.) Covers all species currently known in 12 genera, with descriptive notes for each genus and species, and numerous colour photographs. Introductory chapters cover cultivation, propagation and other topics.
11 Australian Plants Society Maroondah Inc. (formerly Society for Growing Australian Plants Maroondah Inc.) *Flora of Melbourne. A Guide to the Indigenous Plants of the Greater Melbourne Area*. (Hyland House, South Melbourne, 3rd edn, 2001.) Introductory sections on plant communities, soils, regeneration, wetlands, seed collection, propagation, etc., followed by comprehensive guide to more than 1200 native species with brief descriptions and line drawings, and 74 small colour photographs. No keys.
12 Backhouse, G., and Jeanes, J., *The Orchids of Victoria*. (Melbourne Univ. Press, Carlton, 1995.) Comprehensive and authoritative treatment with introductory sections followed by keys and detailed descriptions of all then-known species, with colour photograph and distribution map for each.
13 Bailey, L. H., *Manual of Cultivated Plants*. (Macmillan, New York, rev. edn, 1949.) An excellent botanical handbook with keys and descriptions of families, genera and species cultivated in North America. Illustrations accompany family descriptions.
14 Bailey, L. H., and Bailey, E. Z., *Hortus Third. A Concise Dictionary of Plants Cultivated in the United States and Canada*. (Macmillan Publishing Co., New York, revised & expanded edn, 1976.) Encyclopaedic listing of a large number of plants with brief descriptions, and numerous entries on more general topics.
15 Baines, J. A., *Australian Plant Genera*. (Soc. for Growing Aust. Plants, 1981.) A dictionary of genera providing pronunciation, numbers of species, distribution notes and some references.
16 Bates, R. J., and Weber, J. Z., *Orchids of South Australia*. (Govt Printer, SA, 1990.) Comprehensive text with introductory sections, illustrated keys, descriptions, distribution maps, numerous colour plates, glossary, bibliography.

Bibliography

17 Beadle, N. C. W., *Students Flora of North Eastern New South Wales*. Parts 1–5. (Univ. of New England, Armidale, 1971–84.) Descriptions of families and genera, keys, illustrations.
18 Beadle, N. C. W., Evans, O. D. and Carolin, R. C., *Flora of the Sydney Region*. (Reed, Sydney, rev. edn, 1972.) Brief descriptions of families and genera, detailed keys to species, many illustrations. See later edn listed under Carolin (no. 42).
19 Benson, D. and Howell, J., *Taken for Granted: The Bushland of Sydney and its Suburbs*. (Kangaroo Press in association with Royal Botanic Gardens, Sydney, 1990.) General account of Sydney's flora with many colour photographs. Topics include Sydney's landscape, plants in Aboriginal life, vegetation types, European impact and an appendix listing common species.
20 Benson, D. H., and Melrose, S. C., 'Floristic lists of NSW (IV)'. *Cunninghamia* 3(1): 167–213. An annotated bibliography of 202 floristic lists, mainly from the period 1986–92, relating to various parts of NSW.
21 Benson, D., et al., *Mountain Devil to Mangrove. A Guide to the Natural Vegetation of the Hawkesbury–Nepean Catchment*. (Royal Botanic Gardens, Sydney, 1996.) Slim A4 volume discussing vegetation types not individual genera or species. Includes numerous maps, a species list and 8 pages of colour photographs.
22 Bishop, A., *Field Guide to the Orchids of New South Wales and Victoria*. (Univ. NSW Press, Sydney, 1996.) Compact but comprehensive work with more than 550 species described, and illustrated by colour photographs, with keys, distribution notes, distinguishing features. 2nd edn, 2000.
23 Black, J. M., *Flora of South Australia*. Part 1. (Govt Printer, Adelaide, 3rd edn, 1978.) Ferns, gymnosperms and monocotyledons only. Keys, descriptions and many small line drawings.
24 Black, J. M., *Flora of South Australia*. Parts 1–4. (Govt Printer, Adelaide, 2nd edn, 1943–57.) Keys and descriptions to most South Australian species, many small line drawings. Note supplement listed under Eichler (no. 76). See also no. 127.
25 Blackall, W. E., and Grieve, B. J., *How to Know Western Australian Wildflowers*. Vols 1–4. (Univ. of Western Australia Press, 1954–75.) A large key to species with numerous small illustrations. See also Grieve (no. 100).
26 Blake, T. L., *A Guide to Darwinia and Homoranthus*. (Soc. for Growing Aust. Plants, Maroondah Group, Ringwood, Vic., 1981.) Descriptions, line drawings and some colour photographs.
27 Blomberry, A. M., and Maloney, B., *Proteaceae of the Sydney Region*. (Kangaroo Press, Kenthurst, 1992.) Good descriptions, colour photographs of more than 90 species in 14 genera. Introductory sections cover geology and floral structure.
28 Blood, K., *Environmental Weeds. A Field Guide for SE Australia*. (C. H. Jerram and Assoc., Mt Waverley, 2001.) Covers over 175 species with good colour photographs, brief descriptions, and notes on ecology, means of spread, similar species.
29 Boland, D. J., et al., *Forest Trees of Australia*. (CSIRO/Nelson, Melbourne, 4th edn, 1984.) Revised and greatly enlarged, first pub. 1957. Over 220 species described in detail, well illustrated.

Bibliography

30 Boomsma, C. D., *Native Trees of South Australia*. (Woods and Forests Dept, S.A., 2nd edn, 1981.) Keys, descriptions and drawings.
31 Borror, D. J., *Dictionary of Word Roots and Combining Forms*. (Mayfield Publ. Co., Mountain View, Calif., 1960, repr. 1971.) A useful list of the meanings of Latin and Greek roots, with explanatory notes.
32 Botany School, Univ. of Melbourne, *The Families and Genera of Victorian Plants*. (Bot. School, Univ. of Melb., 11th edn, 1987.) Keys to Victorian dicot and monocot families and genera, descriptions of families, and useful notes on inflorescences, leaf terms, fruits, etc. Edn 12a (1989) partly revised, credits authorship to S. L. Duigan. Edn 14 publ. 1992.
33 Bowes, B. G., *A Colour Atlas of Plant Structure*. (Manson Publishing Ltd, London, 1996.) Includes excellent, extensively labelled colour photographs of plant parts from cells to flowers and leaves etc., with many close-ups and sections with detailed supplementary captions.
34 Brock, J., *Top End Native Plants*. (Author publ., Darwin, 1988.) Brief descriptions of 450 species, with 700 good colour photographs, 26 line drawings, notes on uses, habitat etc.
35 Brooker, M. I. H., 'A New Classification of the genus *Eucalyptus*'. *Australian Systematic Botany* 13: 79–148 (2000). A research paper proposing the inclusion of the genus *Angophora* within an expanded concept of *Eucalyptus*, and setting out a detailed infrageneric classification.
36 Brooker, M. I. H., and Kleinig, D. A., *Eucalypts. An Illustrated Guide to Identification*. (Reed Books, Port Melbourne, 1996.) A selection of 200 'most common spp.' from all over Australia, with colour photographs, brief descriptions, distribution maps, notes. An expanded, illustrated key to groups of species facilitates identification.
37 Brooker, M. I. H., and Kleinig, D. A., *Field Guide to Eucalypts*, vol. 1, *South-eastern Australia*, vol. 2, *South-western and Southern Australia*, vol. 3, *Northern Australia*. (Inkata, Melbourne and Sydney, 1983, 1990, 1994.) Includes a detailed discussion of the important characteristics of the genus, keys, descriptions and colour photographs of all species then known in the regions. Revised edns of vols 1 and 2 published by Bloomings Books, Hawthorn, 1999, 2001.
38 Brown, A., Keighery, G., and Thomson, C., *Common Wildflowers of the South-west Forests*. (Dept Conservation and Land Management, Como, WA, 1996.)
39 Buchanan, A. M., (ed.), *A Census of the Vascular Plants of Tasmania and Index to the Student's Flora of Tasmania*. (Tas. Museum and Art Gallery, Hobart, 1995.) Tasmanian Herbarium Occasional Publication No. 5.
40 Burbidge, N. T., and Gray, M., *Flora of the Australian Capital Territory*. (Australian National Univ. Press, Canberra, 1970.) Amplified keys to species, numerous line drawings.
41 Cameron, M., (ed.), *Guide to Flowers and Plants of Tasmania*. (Reed, Sydney, 1981.) 300 good colour photographs and brief descriptions. Reprinted 1994.
42 Carolin, R. C., and Tindale, M. D., *Flora of the Sydney Region*. (Reed, Chatswood, rev. 4th edn, 1994.) A complete revision of Beadle *et al*. (no 18), with keys and

descriptions for all then-known species in the region, including 16 pages of colour photographs, and a number of pages of line drawings.
43 Carr, D. J., and Carr, S. G. M. (eds), *People and Plants in Australia*. (Academic Press, Sydney, 1981.)
44 Carr, D. J., and Carr, S. G. M. (eds), *Plants and Man in Australia*. (Academic Press, Sydney, 1981.)
45 Chippendale, G. M., 'Checklist of Northern Territory Plants', *Proc. Linn. Soc. NSW* 96 (4): 207–67 (1987). As well as the checklist, includes a descriptive account, and bibliography of works related to the NT flora.
46 Chippendale, G. M., (ed.), *Eucalyptus Buds and Fruits*. (Forestry and Timber Bureau, Canberra, 1968.) Line drawings of buds and fruits of nearly all species then recognised.
47 Clements, M. A., 'Catalogue of Australian Orchidaceae'. *Australian Orchid Research*. Vol. 1. (Australian Orchid Foundation, 1989.) A thorough list of orchid names and synonyms with details of type specimens etc. Attempts to set a benchmark for Australian orchid nomenclature.
48 Clifford, H. T., and Ludlow, G., *Keys to the Families and Genera of Queensland Flowering Plants* (Magnoliophyta). (Univ. of Qld Press, 1972.)
49 Clyne, D., *Australian Ground Orchids*. (Lansdowne Press, Melbourne, 1970.) Brief explanation of the orchid flower, more than 70 colour photographs.
50 Cochrane, G. R., Fuhrer, B. A., Rotherham, E. R., and Willis, J. H., *Flowers and Plants of Victoria*. (Reed, Sydney, rev. edn, 1973.) Over 540 excellent colour photographs arranged according to plant habitats, with brief descriptive text. The 3rd edn expanded, under the title *Flowers and Plants of Victoria and Tasmania*. (1980).
51 Codd, P., et al., *The Plant Life of Kosciuszko*. (Kangaroo Press, East Roseville, NSW, 1998.) A useful guide. Contains descriptions of habitats and communities illustrated with good colour photographs. At least half the book devoted to individual species with descriptions and numerous line drawings.
52 Cook, C. D. K., *Aquatic Plant Book*. (SPB Academic Publishing, The Hague, 1990.) Descriptions of 407 genera, with keys. Well illustrated with line drawings.
53 Cooper, W., and Cooper, W. T., *Fruits of the Rainforest*. (GEO Productions, Chatswood, 1994.) Larger format volume of excellent watercolour illustrations of fruits of 626 species, with brief descriptions, occurring in the tropical rainforests of Queensland.
54 Corrick, M. G., 'Bush-peas of Victoria—genus *Pultenaea*, Nos 1–24', *Victorian Naturalist* vol. 93–107 (1976–90). An excellent series with descriptions and illustrations. The last article includes a key.
55 Corrick, M. G., and Fuhrer, B. A., *Wildflowers of Southern Western Australia*. (The Five Mile Press, Noble Park, Vic., with Monash Univ., 1996.) Brief introductory section, excellent colour photographs of over 700 species, each with brief description, arranged in families. Updated edn published 2002.
56 Corrick, M. G., and Fuhrer, B. A., *Wildflowers of Victoria and Adjoining Areas*. (Bloomings Books, Hawthorn, 2000.) Brief introductory pages and small glossary followed by 838 excellent colour photographs, several per page, each with brief descriptive text, arranged in families.

57 Costermans, L. F., *Native Trees and Shrubs of South-eastern Australia.* (Rigby, Melbourne, 1981.) An excellent guide designed for identification, with species arranged in families, as well as regional guide lists, and informative introductory chapters. Numerous colour photographs and line drawings. Revised edn, Lansdowne Publications, 1994.
58 Costermans, L. F., *Trees of Victoria.* (Pub. by the author, Melbourne, 1973.) Illustrations and descriptions of the common species. A very good field guide. 5th edn published by Costermans Publishing, Frankston, 1994, under the title *Trees of Victoria and Adjoining Areas*.
59 Costin, A. B., *et al., Kosciusko Alpine Flora.* (CSIRO/Collins, 1979.) An authoritative work with keys, descriptions and excellent photographs. Second edn, and an abridged field edn, published in 2000.
60 Cousins, S. W., *Checklist of the Vascular Plants of the Darwin Region, Northern Territory, Australia.* (Conservation Commission of the Northern Territory, 1989.)
61 Cowie, I. D., *et al., Floodplain Flora. A Flora of the Coastal Floodplains of the Northern Territory, Australia.* (Flora of Australia supplementary series, No. 10, ABRS, Canberra/Parks and Wildlife Commission of the NT, Darwin, 2000.) Introductory sections on climate, vegetation, Aboriginal uses, fauna, etc., followed by good line drawings, descriptions and keys for more than 300 species, with 14 pages of colour plates.
62 Craig, G. F., *et al., Native plants of the Ravensthorpe Region.* (Ravensthorpe Wildflower Show Inc., Ravensthorpe, WA, 1995.) Brief introduction, mostly good colour photographs of 161 species with very brief identification notes.
63 Cullen, J., *The Identification of Flowering Plant Families.* (Cambridge Univ. Press, 4th edn, 1997.) Brief explanatory section followed by a key to families.
64 Cunningham, G. M., Mulham, W. E., Milthorpe, P. L., and Leigh, J. H., *Plants of Western New South Wales.* (Govt Printer, Sydney, 1981.) A comprehensive work of many colour photographs, line drawings and descriptions. Later edn published by Inkata, 1993.
65 Curtis, W. M., *The Student's Flora of Tasmania.* Parts 1–4A. (Govt Printer, Tasmania, 1956–80.) Keys and detailed descriptions, many line drawings. Part 1 rev. 2nd edn (1975) and part 4B (St David's Park Publishing, Hobart, 1994) authored by Curtis, W. M., and Morris, D. I.
66 Dashorst, G. R. M., and Jessop, J. P., *Plants of the Adelaide Plains and Hills.* (Kangaroo Press, Kenthurst, NSW, 1990.) Good, partially coloured pencil and watercolour drawings illustrate about 1200 species in nearly 100 plates. Includes brief descriptive text and map of distribution in the Southern Lofty region for each species.
67 Debenham, C., *The Language of Botany.* (Soc. for Growing Aust. Plants, no date.)
68 Debenham, C. N., 'The Family Caesalpiniaceae', *Australian Plants*, 8: 361–4 (1976).
69 Debenham, C. N., 'The Plant Family Fabaceae', *Australian Plants*, 10: 124–39 (1979).
70 Dockrill, A. W., *Australian Indigenous Orchids*, vol. 1. (Soc. for Growing Aust. Plants, 1969.) A large text dealing with epiphytes and tropical terrestrial orchids. Over 300 species and forms illustrated and described in detail. Revised edn published by Surrey Beatty and Sons, Chipping Norton, 1992. Includes keys and some colour photographs.

Bibliography

71 Duigan, S. L., *The Grasses, Particularly those of Victoria, Australia*. (Botany School, Univ. of Melbourne, 1987, repr. 1991.) A detailed, illustrated account, including a key to Victorian genera.
72 Dunlop, C. R., *et al., Checklist of the Vascular Plants of the Northern Territory, Australia*. (Conservation Commission of the NT, Darwin, 1995.)
73 Dunlop, C. R., *et al., Flora of the Darwin Region*. (Conservation Commission of the Northern Territory, Palmerston, vol. 2, 1995.) Comprehensive and authoritative, vol. 2 (the first published of 4 vols) covers 440 species of the total of about 1500, including legumes. Includes keys, descriptions and line drawings.
74 Eddy, D., *et al., Grassland Flora. A Field Guide for the Southern Tablelands*. (World Wide Fund for Nature, Australian National Botanic Gardens, Canberra, and others, 1998.) Small-format, spiral-bound text contains good-quality colour photographs (several per page) arranged by life forms, with very brief descriptive notes.
75 Edmonds, T., and Webb, J., *Sydney Sandstone Flora*, (NSW Univ. Press, Kensington, 1986.) A small and useful introduction to plant structure and some common families of the area. Revised edn published under the title *Sydney Flora. A Beginner's Guide to Native Plants* by Surrey Beatty and Sons, Chipping Norton, NSW, 1998.
76 Eichler, H. J., *Supplement to the Flora of South Australia*. (Govt Printer, Adelaide, 1965.) Brings nomenclature of Black's Flora (no. 24) up to date to 1965.
77 Elliot, R., *A Field Guide to the Grampians Flora*. (Algona, Melbourne, 1975, rev. edn, 1984.) A very useful, well-illustrated, small-format guide.
78 Elliot, W. R., and Jones, D. L., *Encyclopaedia of Australian Plants Suitable for Cultivation*. Vols 1– . (Lothian, Melbourne, 1980– .) A comprehensive, well-illustrated work, with 8 vols published to date (genera A–So), plus 3 supplements in loose-leaf format. Notes on families, brief descriptions of genera and species.
79 Erickson, R., George, A. S., Marchant, N. G., and Morcombe, M. K., *Flowers and Plants of Western Australia*. (Reed, Sydney, 1973.) Over 530 excellent photographs with brief descriptive text.
80 Erickson, R., *Plants of Prey*. (Lamb Publications, WA, 1968; and Univ. of WA Press, 1975.) Well illustrated, comprehensive at the time.
81 Erickson, R., *Triggerplants*. (Paterson Brokensha, Perth, 1958.) Well illustrated, comprehensive in its time.
82 Fairley, A., and Moore, P., *Native Plants of the Sydney District. An Identification Guide*. (Kangaroo Press, Kenthurst, in assoc. with SGAP-NSW, 1989.) Comprehensive guide with nearly 1500 species of dicots, monocots (including grasses and sedges), ferns and gymnosperms arranged in families, described and illustrated with good colour photographs. Includes identification tables ('keys'), references, glossary.
83 Floyd, A. G., *Key to Major Rainforest Trees in N.S.W.* (Forestry Commission of NSW, Sydney, 2nd edn, 1977.) Research note No. 27.
84 Floyd, A. G., *N.S.W. Rainforest Trees*. (Forestry Commission of NSW, Sydney, 1961– .) A series of research notes, many in the 2nd edn, including nos 7, 28, 29, 32, 34, 35, 38, 41, 43, 48, 49. Keys and descriptions.

Bibliography

85 Floyd, A. G., *Rainforest Trees of Mainland South-eastern Australia*. (Inkata, Melbourne, 1989.) A comprehensive and authoritative work including descriptions and illustrations of 385 species, with keys. Based on nos 83 and 84.

86 Forbes, S. J., 'Notes from the National Herbarium of Victoria—1, Review of recent studies in *Eucalyptus*—the genus; —2, Review of recent studies in *Eucalyptus*—the species', *Victorian Naturalist*, 103: 183–4 (1986); 104: 14–19 (1987). A brief but informative review with references.

87 Forbes, S. J., Gullan, P. K., Kilgour, R. A., and Powell, M. A., *A Census of the Vascular Plants of Victoria*. (National Herbarium, Dept Conservation, Forests and Lands, Melbourne, 1984.) An alphabetical list of then-current names of families, genera and species. The later 5th edn was the last to indicate names current in Willis' handbook (no. 233). Most later edns, including the 6th (2000), edited by Ross, J. H.

88 Foster, E., and MacDonald, M., *Orchids of the Anglesea District*. (Inverted Logic, Brunswick, Vic., 1999.) Brief introduction plus 100 species illustrated by colour photographs, with brief notes.

89 Francis, W. D., *Australian Rain-forest Trees*. (Aust. Govt Pub. Service, Canberra, 4th edn, 1981.) Includes a key and detailed descriptions. Many black-and-white photographs.

90 Fuller, L., *Wollongong's Native Trees*. (Author publ., Kiama, 1980.) Contains introductory sections followed by some 250 pages (two per species) of useful black-and-white photographs with brief descriptions and notes.

91 Galbraith, J., *Collins Field Guide to the Wildflowers of Southeast Australia*. (Collins, Sydney, 1977.)

92 George, A. S., *An Introduction to the Proteaceae of Western Australia*. (Kangaroo Press, Kenthurst, NSW, 1984.) Sixteen genera introduced with one or more species then described and illustrated, with 162 colour photographs.

93 George, A. S., *The Banksia Book*. (Kangaroo Press and Soc. for Growing Aust. Plants, NSW Ltd, Kenthurst, 3rd edn, 1996.) Detailed, thorough, including many colour photographs.

94 George, A. S., (ed.), *Flora of Australia*. Vols 1– . (Aust. Govt Publ. Service, Canberra, 1981– .) A large series of volumes including keys and descriptions to all Australian species known at the time of publication, with numerous illustrations (line drawings and colour photographs). Some 20 vols published to date covering nearly 100 families of flowering plants (ferns, gymnosperms and allies are treated in vol. 48). Editors of later volumes (after 1992) include Orchard, A. E., Thompson H. S., and McCarthy, P. M. The 2nd edn of the introductory vol. 1, considerably enlarged, was edited by Orchard, A. E., (1999). Both edns include a chapter on classification. A complete flora of Norfolk I. and Lord Howe I. is presented in vol. 49. Other offshore islands and territories are dealt with in vol. 50. Families/genera (with volume number) covered include *Acacia* 11A, 11B, Brassicaceae 8, Caesalpiniaceae 12, Casuarinaceae 3, Chenopodiaceae 4, Droseraceae 8, *Eucalyptus* 19, Goodeniaceae 35, *Grevillea* 17A, Iridaceae 46, Liliaceae 45, Loranthaceae 22,

Bibliography

Mimosaceae 11A, 11B, 12, Proteaceae 16, 17A, 17B, Sapindaceae 25, Solanaceae 29, Stackhousiaceae 22, Thymelaeaceae 18, Xanthorrhoeaceae 46.

95 George, E., *Verticordia. The Turner of Hearts.* (Univ. of WA Press, Nedlands, 2002.) Comprehensive guide to 101 species with watercolour illustrations and line drawings as well as numerous colour photographs, location maps, cultivation notes, general interest information.

96 Gledhill, D., *The Names of Plants.* (Cambridge Univ. Press, 2nd edn, 1989.) Provides the meanings of scientific names, with introductory chapters on the history and rules of nomenclature.

97 Gray, C. E., *Victorian Native Orchids.* Vols 1 and 2. (Longman, Melbourne, 1966, 1971.) Black-and-white or colour photographs of nearly all Vic. species then recognised. Very brief descriptions.

98 Green, J. W., *Census of the Vascular Plants of Western Australia.* (WA Herbarium, Dept Agriculture, Perth, 1985.)

99 Greuter, W., *et al.* (eds), *International Code of Botanical Nomenclature.* (Koeltz Scientific Books, Konigstein, Germany, 2000.) The code is also available on the internet—see W4.

100 Grieve, B. J., *How to know Western Australian Wildflowers* Part 4. (Univ. WA Press, Nedlands, 2nd edn, 1998.) Standard text to identification of WA plants. Essentially a large key (covering families Amaranthaceae to Lythraceae in taxonomic sequence) with small line drawings of each species. Includes 14 pages of colour photographs (6 or more per page). See also Blackall and Grieve (no. 25).

101 Hacker, J. B., *A Guide to the Herb and Shrub Legumes of Queensland.* (Univ. Qld Press, St Lucia, 1990.) An illustrated guide including an extensive bibliography.

102 Hall, N., and Brooker, I., *Forest Tree Leaflets.* Vols 1–4. (CSIRO, Melbourne, 1980.) Detailed descriptions with illustrations of some 200 species.

103 Hall, N., and Johnson, L. A. S., *The Names of Acacias of New South Wales with a Guide to Pronunciation of Botanical Names.* (Royal Botanic Gardens, Sydney, 1993.)

104 Harden, G. J. (ed.), *Flora of New South Wales.* Vols 1–4. (Univ. NSW Press, Kensington, 1990–93.) The standard text for identification of all then-known species, large format, with keys, descriptions, small line drawings of all species, plus numerous colour photographs. Supplement to vol. 1 edited by Harden, G. J., and Murray, L. J., (2000).

105 Harden, G., and Williams, J. B., *Fruits—a Guide to Some Common and Unusual Fruits found in the Rainforests.* (Univ. of New England Press, Armidale, 1979.)

106 Harris, K., *Flowers and Ferns of Morwell National Park.* (Friends of Morwell National Park, 1997.) More than 380 colour photographs arranged 12 per page, introductory notes, large species list.

107 Hauser, P. J., *Fragments of Green. An Identification Guide for Rainforest Plants of the Greater Brisbane Area.* (Rainforest Conservation Society, Bardon, 1992.) Includes introductory section, simplified key to groups, glossary etc. then some 360 pages usually with one species covered per page, with brief identification notes and good line drawings.

Bibliography

108 Henderson R. J. F., (ed.) *Queensland Plants: Names and Distribution.* (Qld Herbarium, Dept of Environment, Toowong, 1997.) A census including mosses, lichens and algae as well as vascular plants.

109 Heywood, V. H., (ed.), *Flowering Plants of the World.* (Oxford Univ. Press, Melbourne, 1978.) An excellent introduction to flowering plant families.

110 Hickey, M., and King, C., *100 Families of Flowering Plants.* (Cambridge Univ. Press, Melbourne, 1981.) Botanical descriptions and one detailed illustrated example of each of 100 flowering plant families. A revised version published in 1997 under the title *Common Families of Flowering Plants.*

111 Hill, K. D., and Johnson, L. A. S., 'Systematic studies in the eucalypts. 7. A revision of the bloodwoods, genus *Corymbia* (Myrtaceae)'. *Telopea* 6(2–3): 185–504 (1995). A research paper in which the new genus *Corymbia* is formally published, and 113 species transferred to it from *Eucalyptus.*

112 Hnatiuk, R. J., *Census of Australian Vascular Plants.* (Aust. Govt Publ. Service, Canberra, 1990. Australian Flora and Fauna series no. 11) A large list of then-current names (17,590 spp.) with distribution data.

113 Hoffman, N., and Brown, A., *Orchids of South-west Australia.* (Univ. of WA Press, Nedlands, revised 2nd edn, 1998.) Includes nearly 450 species illustrated in colour, with maps, descriptions, and notes on distinctive features, flowering times and habitats.

114 Holliday, I., *A Field Guide to Melaleucas.* (Hamlyn, Port Melbourne, 1989.) Fairly comprehensive with about 140 species covered. Good colour photographs, many small line drawings, with brief descriptive text. Volume 2, updating vol. 1 and dealing with nearly 50 spp., was published by the author in 1997.

115 Holliday, I., and Watton, G., *A Field Guide to Banksias.* (Rigby, Adelaide, 1975.) Colour photographs, brief descriptions of the majority of species.

116 Holliday, I., and Watton, G., *A Gardener's Guide to Eucalypts.* (Rigby, Adelaide, 1980.) Colour photographs and brief descriptions of over 120 species.

117 Hopper, S. D., *Kangaroo Paws and Catspaws. A Natural History and Field Guide.* (Dept Conservation and Land Management, Como, WA, 1993.) Comprehensive small-format guide, with chapters on natural history topics, relationships, biology, habitats etc., followed by descriptive notes on the species, each well illustrated by several good colour photographs.

118 Hubbard, C. E., *Grasses.* (Penguin, Ringwood, Vic., 3rd edn, 1984.) An excellent introduction. Well illustrated, covering species of the British Isles, many of which are introduced in Australia. Key and detailed descriptions included.

119 Hussey, B. M. J., et al., *Western Weeds. A Guide to the Weeds of Western Australia.* (Plant Protection Soc. WA, Victoria Park, 1997.) Numerous good colour photographs with brief descriptive text.

120 Hyde-Wyatt, B. H., and Morris, D. I., *Tasmanian Weed Handbook. A Guide to the Identification of the Main Broad-leaf Weeds of Crops and Pastures in Tasmania.* (Dept Agriculture, Hobart, 1975.) Descriptions and good line drawings with a key to seedlings.

Bibliography

121 Hyde-Wyatt, B. H., and Morris, D. I., *The Noxious and Secondary Weeds of Tasmania.* (Dept of Agriculture, Hobart, 1980.) More than 100 species illustrated with good line drawings, with accompanying notes including descriptions of seedlings.

122 Hyland, B. P. M., et al., *Australian Tropical Rainforest Trees and Shrubs. An Interactive Identification System for Trees and Shrubs.* (CSIRO Publishing, Collingwood, 1999.) Covers over 1000 species and includes an introductory and explanatory volume (with discussion of plant features, and species list), a volume of plant descriptions, diagnostic features etc., a volume of leaf photographs, and a CD with a computer key to species. A later, electronic only, expanded edn is listed here as CD3.

123 Jacobs, S. W. L., and Pickard, J., *Plants of New South Wales.* (Govt Printer, NSW, 1981.) A census of NSW species.

124 Jessop, J. P., 'Australian Lamiaceae', *Australian Plants*, 10: 299–307 (1980).

125 Jessop, J. P., *A List of the Vascular Plants of South Australia*, (Adelaide Botanic Gardens and State Herbarium, 2nd edn, 1984.) The 4th edn was published in 1993.

126 Jessop, J. P., (ed.), *Flora of Central Australia.* (Reed, Sydney, 1981.) Keys, descriptions and numerous line drawings. Covers a broad area of inland Australia. Reprinted 1985.

127 Jessop, J. P., and Toelken, H. R., (eds), *Flora of South Australia.* Parts 1–4. (Govt Printer, Adelaide, 4th edn, 1986.) A completely revised edition of Black's *Flora* (nos 23, 24). The standard identification text for SA species, with keys, descriptions and numerous line drawings.

128 Johnson, P., and Brooke, P., *Wetland Plants in New Zealand.* (DSIR Publishing, Wellington, 1989.) Includes more than 500 good pencil illustrations, brief descriptions emphasising distinctive features, some colour habitat photographs, distributions and notes.

129 Jones, D. L., *Native Orchids of Australia.* (Reed Books, Frenchs Forest, 1988.) Comprehensive work with all then-known species described (about 700), 260 illustrated in colour, with over 600 line drawings. Includes notes on similar species and cultivation. No keys.

130 Jones, D. L., *Ornamental Rainforest Plants in Australia.* (Reed, Frenchs Forest, 1986.) Comprehensive, with brief descriptions including propagation and cultivation notes of more than 1000 species, over 250 colour photographs and 100 line drawings. A number of appendices list plants showing various features.

131 Jones, D. L., *Palms in Australia.* (Reed, Port Melbourne, 1996.) Comprehensive, well illustrated with good colour photographs, includes descriptions.

132 Jones, D. L., and Gray, B., *Australian Climbing Plants.* (Reed, Sydney, 1977.) Some 250 species illustrated, most in colour, with notes on identification, horticulture etc.

133 Jones, D. L., and Gray, B., *Climbing Plants in Australia with emphasis on Australian Native Species.* (Reed Books, Frenchs Forest, 1988.) Some 750 species included, both native and exotic, with brief descriptions, many colour photographs, introductory section. Leaf outline sketches of 100 spp. act as identification aid. (A revised and enlarged edition of no. 132.)

134 Jones, D. L., and Jones, B., *A Field Guide to the Native Orchids of Southern Australia.* (Bloomings Books, Hawthorn, 2000.) Most species in the region are covered one

per page with colour photographs, and brief descriptions, distribution, similar species and habitat notes.
135 Jones, D. L., et al., *The Orchids of Tasmania*. (Melbourne Univ. Press, Carlton, 1999.) A comprehensive work with keys to genera and species, introductory sections on habitats, conservation, structure etc. followed by one species per page described and illustrated with colour photographs.
136 Jones, S. B., and Luchsinger, A. E., *Plant Systematics*. (McGraw-Hill, New York, 1979.) A standard botany textbook containing a useful chapter on the origin and classification of flowering plants.
137 Kelly, S., *Eucalypts*. Vols 1 and 2. (Nelson, Melbourne, 1969, 1978.) Watercolour illustrations of almost all species then known with descriptive text.
138 Kirkpatrick, J., *Alpine Tasmania. An Illustrated Guide to the Flora and Vegetation*. (Oxford Univ. Press, Melbourne, 1997.) Most of the book comprises brief but useful descriptions and numerous accompanying line drawings of the vascular plants of the region (more than 400 spp.), grouped according to life form. Includes introductory sections plus 16 pages of good colour photographs.
139 Kleinschmidt, H. E., and Johnson, R. W., *Weeds of Queensland*. (Qld Dept Primary Industries, Brisbane, 1987.) Most of the book deals with 369 species, one per page, each with brief description and black-and-white photograph. Includes identification guide to groups illustrated with 16 pages of colour photographs.
140 Kubitzki, K., et al. (eds), *The Families and Genera of Vascular Plants*. Vols 2–4. (Sprenger Verlag, Berlin, 1993–98). Comprehensive and scholarly discussion of families, including keys to genera. Vol. 2 includes 73 dicot families, vols 3 and 4 deal with monocots. (Vol. 1 deals with ferns and gymnosperms.)
141 Lamp, C. A., et al., *Grasses of Temperate Australia. A Field Guide*. (Inkata Press, Melbourne, 1990, 2nd edn, 2001.) Mostly comprises descriptions and illustrations (quality of reproduction not always good) of more than 100 species. Includes introductory sections on structure, importance, troublesome features, grassland, and an illustrated aid to identification.
142 Lamp, C., and Collet, F., *A Field Guide to Weeds in Australia*. (Inkata Press, Melbourne, rev. edn 1984.) Colour photographs with very brief descriptive text.
143 Lawrence, G. H. M., *Taxonomy of Vascular Plants*. (Macmillan, New York, 1951.) A standard botany text, including detailed chapters on the structure of plants, descriptions of families, glossary.
144 Lazarides, M., *The Grasses of Central Australia*. (Aust. Nat. Univ. Press, Canberra, 1970.) Keys, descriptions, black-and-white photographs.
145 Lear, R., and Turner, T., *Mangroves of Australia*. (Univ. of Qld Press, St Lucia, 1977.) Small format, including chapters on habitat, plant characteristics, etc., notes on 8 main species, and field key to Australian species.
146 Leonard, G., *Eucalypts. A Bushwalker's Guide*. (NSW Univ. Press, Kensington, 1993.) Pocket-sized, well-illustrated guide to 68 species from the bush around Sydney, each with brief description, line drawing of buds and fruits, and colour picture of bark. Key to groups of species based on bark features.

Bibliography

147 Little, R. J., and Jones, C. E., *A Dictionary of Botany*. (Van Nostrand, Melbourne, 1980.)
148 Lovelock, C., *Field Guide to the Mangroves of Queensland*. (Aust. Inst. Marine Science, 1993.) Well-presented guide to 22 species, with identification notes, line sketches of habit and colour paintings of leafy branches.
149 Lowrie, A., *Carnivorous Plants of Australia*. Vols 1–3. (Univ. WA Press, Nedlands, 1987, 1989, 1998.) Comprehensive coverage with each species described and very well illustrated with good photographs and line drawings, with descriptions and distribution maps. Includes keys.
150 Lumley, P. F., and Spencer, R. D., *Plant Names*. (Royal Botanic Gardens, South Yarra, 2nd edn 1991, reprinted 1995.) A very useful introduction, includes sections on name construction, common names, the International Code of Botanical Nomenclature (including that for cultivated plants), name changes, and pronunciation.
151 Lunt, I., *et al.*, *Plains Wandering. Exploring the Grassy Plains of South Eastern Australia*. (Vict. Nat. Parks Assoc. and Trust for Nature, 1998.) 38 pages of introduction, ecology and regional descriptions, plus more than 320 photographs covering some 280 species each with brief descriptive sentence.
152 Mabberley, D. J., *The Plant-book. A Portable Dictionary of the Vascular Plants*. (Cambridge Univ. Press, Cambridge, 2nd edn, 1997.) A useful compact dictionary of families and genera, providing brief information on taxonomy and distribution etc.
153 Marchant, N. G., *et al.*, *Flora of the Perth Region* Parts 1–2. (Western Australian Herbarium, Perth, 1987.) Comprehensive and authoritative, with keys and descriptions for some 2000 species from the region. Numerous line drawings.
154 Marriott, N. and Marriott, J., *Grassland Plants of South-eastern Australia*. (Bloomings Books, Hawthorn, 1998.) About 170 colour photographs, one per page with brief descriptive text.
155 McCann, I. R., *The Alps in Flower*. (Victorian National Parks Association, East Melbourne, 3rd edn, 1996.) About 60 pages of good colour photographs (three or four per page) with brief descriptive captions.
156 McCann, I. R., *The Coast and Hinterland in Flower*. (Victorian National Parks Association, East Melbourne, 1992.) Over 100 pages of good colour photographs (three or four per page) with brief descriptive captions.
157 McCann, I. R., *The Grampians in Flower*. (Victorian National Parks Association, East Melbourne, 1994.) Over 100 pages of good colour photographs (three or four per page) with brief descriptive captions.
158 McCann, I. R., *The Mallee in Flower*. (Vict. Nat. Parks Assoc., E. Melbourne, 1989.) Three or four good colour photographs on each of 120 pages, with brief descriptive captions.
159 McGillivray, D. J., *Grevillea*. (Melbourne Univ. Press, Carlton, 1993.) A very large and comprehensive study of the whole genus as it was then known. Includes keys, descriptions, maps, drawings, numerous colour photographs.
160 McMahon, D. S., *Eucalypts for Enthusiasts. A Guide to the Identification of Eucalypts found in South-eastern Australia*. (Author publ., Fitzroy, 1990.) Small-format, useful

guide, set out in key form throughout, to all then-recognised species in the region.
161 Meney, K. A., and Pate, J. S. (eds), *Australian Rushes. Biology, Identification and Conservation of Restionaceae and Allied Families*. (Univ. of WA Press, Nedlands, 1999.) Comprehensive and authoritative work, includes excellent illustrations, keys and descriptions.
162 Mitchell, A., *et al.*, *Arid Shrubland Plants of Western Australia*. (Univ. of WA Press, Nedlands, revised edn, 1999.) Includes about 200 species illustrated in colour, with notes and brief descriptions.
163 Moerkerk, M., and Barnett, A. G., *More Crop Weeds* (R. G. & F. J. Richardson, Meredith, 1998.) Briefly describes 90 weeds, with more than 300 colour, and 90 black-and-white photographs illustrating various stages of growth. See also no. 229.
164 Morley, B. D., and Toelken, H. R. (eds), *Flowering Plants in Australia*. (Rigby, Adelaide, 1983.) A large, well-illustrated, and informative work on plant families, including keys to the genera in each family.
165 Morris, D. I., *Grasses of Tasmania*. (Tasmanian Herbarium, Hobart, Occasional Paper No. 3, 1991.) Keys and descriptions to all then-known Tas. species.
166 Muyt, A., *Bush Invaders of South-East Australia*. (R. G. & F. J. Richardson, Meredith, 2001.) A major part of this book deals with the problems of environmental weeds and their management and control. The second section deals with 150 species in 93 weed descriptions, with colour photographs and notes on diagnostic features, similar indigenous species etc.
167 Nicholls, W. H., *Orchids of Australia*. (Nelson, Melbourne, 1969.) Watercolour paintings of some 470 species in great detail, with descriptive text.
168 Nicholson, N., and Nicholson, H., *Australian Rainforest Plants*. Vols 1–5. (H. & N. Nicholson, Terrania Rainforest Publishing, The Channon, NSW, vol. 1, 4th edn, 1992; vol. 2, 2nd edn, 1991; vol. 3, 1991; vol. 4, 1994; vol. 5, 2002.) Small books each with good colour photographs of over 100 species, with brief descriptions and garden notes.
169 Olde, P., and Marriott, N., *The Grevillea Book*. Vols 1–3. (Kangaroo Press, Kenthurst, vol. 1, 1994, vols 2 and 3, 1995.) Comprehensive, detailed and very well-illustrated work on all then-known species, as well as numerous forms in cultivation. Includes keys, descriptions and usually several good colour photographs per species.
170 Paczkowska, G., and Chapman, A. R., *The Western Australian Flora. A Descriptive Catalogue*. (Wildflower Soc. of Western Australia Inc., The Western Australian Herbarium, and the Botanic Gardens and Parks Authority, Perth, 2000.) Comprehensive listing of 11,922 plants arranged alphabetically by families etc., each with a brief descriptive statement noting habit, height, flowering time and colour, habitat and distribution. Includes extensive bibliography and appendices.
171 Parsons, M. J., *et al.* (eds), *Current Names List for Wild Gymnosperms, Dicotyledons and Monocotyledons (except Grasses) in New Zealand as used in Herbarium CHR*. (Manaaki Whenua Press, Canterbury, New Zealand, 1995.) List of names with basic bibliographic details, coded distributions, etc.

Bibliography

172 Parsons, W. T., *Noxious Weeds of Victoria*. (Inkata, Melbourne, 1973.) Informative and well illustrated.
173 Parsons, W. T., and Cuthbertson, E. G., *Noxious Weeds of Australia*. (Inkata Press, Melbourne, 1992.) Comprehensive and authoritative, detailed descriptions, numerous colour photographs, distribution maps, no keys. Rev. edn published by CSIRO in 2000.
174 Pate, J. S., and McComb, A. J. (eds), *The Biology of Australian Plants*. (Univ. of Western Australia Press, Nedlands, 1981.) A series of research papers on aspects of the biology of Australian plants, e.g., responses to fire, salinity, water stress, nutrient deficient soils etc., as well as sections on eucalypt forests, seagrasses, resurrection plants, arborescent monocots among others.
175 Pedley, L., *Acacias in Queensland*. (Qld Herbarium, Dept Primary Industries, Indooroopilly, 1991.) Comprehensive descriptions of Qld species, with keys.
176 Petheram, R. J., and Kok, B., *Plants of the Kimberley Region of Western Australia*. (Univ. of WA Press, Nedlands, 1983.) Brief introductory pages followed by 242 species illustrated with colour photographs and briefly described, with notes on occurrence and feed value.
177 Pocock, M. R., *Ground Orchids of Australia*. (Jacaranda Press, Milton, Qld, 1972.)
178 Prescott, A., *It's Blue with Five Petals. Wildflowers of the Adelaide Region*. (Author publ., Prospect, SA, 1988.) A well-set-out and well-illustrated field guide, with more than 1000 line drawings, using a simplified colour-coded identification procedure. Includes weeds but grasses, sedges and rushes omitted.
179 Pryor, L. D., and Johnson, L. A. S., *A Classification of the Eucalypts*. (Australian National Univ., Canberra, 1971.) A landmark research publication in the history of eucalypt classification in which an informal subdivision of the genus is proposed. Contains an informative discussion of characteristics of the genus and of previous work on its classification.
180 Randell, B. R., and Symon, D. E., 'The Species of *Cassia* in Australia', *Australian Plants*, 8: 345–60 (1976).
181 Raven, P. H., et al., *Biology of Plants*. (Worth Publishers, New York, 4th edn, 1986.) A standard botany textbook, very well illustrated, with useful chapters on classification and the evolution of flowering plants.
182 Rendle, A. B., *The Classification of Flowering Plants*. Vol. 1 Monocotyledons, vol. 2 Dicotyledons. (Cambridge Univ. Press, London, 1930, 1938, repr. 1952–53.) An older text but useful for the detailed discussions of families.
183 Rippey, E., and Rowland, B., *Plants of the Perth Coast and Islands*. (Univ. WA Press, Nedlands, 1995.) Introductory chapters on climate, topography etc., brief descriptions and good watercolour sketches of the plants with distribution maps and notes.
184 Robinson, L., *Field Guide to the Native Plants of Sydney*. (Kangaroo Press, Kenthurst, 1991.) Comprehensive small-format text covers about 1370 species arranged in families, illustrated with small line drawings, with brief descriptions, general notes, as well as a field key to families and main groups.
185 Rogers, F. J. C., *A Field Guide to Victorian Wattles*. (Pub. by the author, rev. edn, 1978.) Useful, small-format introduction, with very brief descriptions, line drawings,

Bibliography

and key to groups of species aiding identification. 3rd edn published by La Trobe Univ. Press, 1993, repr. 1995.

186 Romanowski, N., *Aquatic and Wetland Plants. A Field Guide for Non-tropical Australia.* (Univ. NSW Press, Sydney, 1998.) More than 340 species briefly described and illustrated with colour photographs (about 220) and drawings.

187 Rotherham, E. R., Blaxell, D. F., Briggs, B. G., and Carolin, R. C., *Flowers and Plants of New South Wales and Southern Queensland.* (Reed, Sydney, 1975.) Over 550 excellent photographs with brief descriptive text.

188 Rowell, R. J., *Ornamental Flowering Shrubs in Australia.* (Univ. NSW Press, Kensington, 2nd edn, 1991.) Brief introductory sections followed by extensive listing of species (alphabetically by genus) each with useful description, horticultural note, and list of cultivars. Includes 28 pages of colour photographs with up to about 8 per page.

189 Rowell, R. J., *Ornamental Flowering Trees in Australia.* (Univ. NSW Press, Kensington, 2nd edn, 1991.) Brief introductory sections followed by extensive listing of species (alphabetically by genus) each with useful description, horticultural note and list of cultivars. Includes 28 pages of colour photographs with up to about 8 per page.

190 Rowell, R. J., *Ornamental Plants for Australia.* (Univ. NSW Press, Kensington, 3rd edn, 1986.) Useful treatment of mostly herbaceous species grouped into annuals, soft-wooded perennials, bulbous plants and climbers (arranged alphabetically by genus) each with brief description, horticultural notes etc. Includes 32 pages of colour photographs with up to about 4 per page.

191 Roy, B., *et al.* (eds), *An Illustrated Guide to Common Weeds of New Zealand.* (Plant Protection Soc. of New Zealand, 1998.) Descriptions of over 280 species, most with colour photograph. Includes key based on flower colour.

192 Rupp, H. M. R., *The Orchids of New South Wales.* Facsimile edn, with a supplement by D. J. McGillivray. (Govt Printer, Sydney, 1969.)

193 Sainsbury, R. M., *A Field Guide to* Dryandra. (Univ. WA Press, Nedlands, 1985.) Comprehensive guide to all then-named species (about 60) but many others still to be named. Several colour photographs per species with brief descriptions and distribution maps.

194 Sainsbury, R. M., *A Field Guide to Isopogons and Petrophiles.* (Univ. WA Press, Nedlands, 1987.) Colour photographs and brief descriptions of more than 60 species.

195 Sainsbury, R. M., *A Field Guide to Smokebushes and Honeysuckles* (Conospermum *and* Lambertia). (Univ. WA Press, Nedlands, 1991.) Colour photographs of some 40 species, with brief descriptions and distribution maps. Concentrates on WA species.

196 Sainty, G. R., and Jacobs, S. W. L., *Waterplants in Australia. A Field Guide.* (Sainty and Assoc., Darlinghurst, 3rd edn, 1994.) Small-format, extensively illustrated, with small pictorial key.

197 Sainty, G., *et al., Alps Invaders. Weeds of the Australian High Country.* (Australian Alps Liaison Committee, 1998. Produced by Sainty and Assoc., Potts Point, NSW.) Small-format, includes a few introductory pages then good colour photographs of 50 species with accompanying notes.

Bibliography

198 Salkin, E., *et al., Australian Brachyscomes*. (Australian Daisy Study Group, Melbourne, 1995.) Some 30 pages of introductory notes (mostly on cultivation) then 74 species described in detail (mostly from cultivated material) and very well illustrated by line drawings, with distribution and cultivation notes.

199 Salmon, J. T., *The Native Trees of New Zealand*. (Reed Methuen, Auckland, rev. edn, 1986.) Some 40 pages of introductory sections then comprehensive coverage of all tree species with more than 1500 excellent colour photographs and brief descriptive text.

200 Scarlett, N. H., *et al., Field Guide to Victoria's Native Grasslands: Native Plants of Victorian Lowland Plains*. (Victoria Press, South Melbourne, 1992.) Introductory and explanatory sections followed by colour photographs of about 120 species with very brief descriptions.

201 Scott, R., *et al., Indigenous Plants of the Sandbelt. A Gardening Guide for South-eastern Melbourne*. (Earthcare, St Kilda, Vic., 2002.) Well presented, with useful and informative introductory sections (60 pp.), followed by brief descriptions and colour photographs of 180 species, plant lists and websites.

202 Semeniuk, V., *et al., Mangroves of Western Australia*. (Western Australian Naturalists Club, Nedlands, 1978.) Handbook No. 12. Keys, descriptions and good line drawings for 17 species in 15 genera. Includes introduction and some colour photographs.

203 Sharpe, P. R., *Keys to Cyperaceae, Restionaceae and Juncaceae in Queensland*. (Queensland Dept Primary Industries, Brisbane, 1986.)

204 Sharr, F. A., *Western Australian Plant Names and their Meanings*. (Univ. of Western Australia Press, Nedlands, 1978, enlarged edn reissued, 1996.) Includes a section on pronunciation of botanical names.

205 Simmons, M. H., *Acacias of Australia*. (Nelson, Melbourne, 1981.) 150 excellent line drawings with descriptions. Some colour plates. Vol. 2 similar in style and content, published by Viking O'Neill, Melbourne, 1988.

206 Simon, B. K., *A Key to Australian Grasses*. (Qld Dept Primary Industries, Brisbane, 2nd edn, 1993.) Brief introductory sections outline classification and grass structure. Keys to genera and species.

207 Simon, B. K., and Latz, P., *A Key to the Grasses of the Northern Territory, Australia*. (Conservation Commission of the Northern Territory, Darwin, 1994.)

208 Society for Growing Australian Plants, SA Region, *Eremophilas for the Garden*. (SGAP, SA Region, 1997.) Nearly 70 colour photographs with brief descriptions and notes. SGAP is now Australian Plants Society.

209 Spencer, R. D., *Horticultural Flora of South-eastern Australia*. Vol. 1, *Ferns, Conifers & their Allies*. Vol. 2, *Flowering Plants, Dicotyledons*, part 1. Vol. 3, *Flowering Plants, Dicotyledons*, part 2. Vol. 4, *Flowering Plants, Dicotyledons*, part 3. (Univ. of NSW Press, Sydney, 1995, 1997, 2002, 2002.) Comprehensive coverage with keys and descriptions for nearly all plants known in cultivation at the time (excluding specialist collections), with small line drawings of most species, and several pages of colour plates. Vol. 5, *Monocotyledons*, in preparation.

Bibliography

210 Stanley, T. D., and Ross, E. M., *Flora of South-eastern Queensland*. Vols 1–3. (Qld Dept Primary Industry, Brisbane, 1983, 1986, 1989.) Keys and descriptions to all species in the region.

211 Stephens, K. M., and Dowling, R. M., *Wetland Plants of Queensland*. (CSIRO Publishing, Collingwood, 2002.) Ninety species covered, one per page with colour photographs, distribution maps and descriptive notes. Includes keys to genera and species.

212 Tame, T., *Acacias of Southeast Australia*. (Kangaroo Press, Kenthurst, 1992.) Detailed descriptions and line drawings of all then-known species in the region. With keys and numerous colour photographs.

213 Thomson, B. G., and Kube, P. D., *Arid Zone Eucalypts of the Northern Territory*. (Conservation Commission of the NT, Alice Springs, 1990.) Descriptions and good line drawings of some 29 species occurring in the southern half of the NT. Includes key, and 28 small colour photographs.

214 Tothill, J. C., and Hacker, J. B., *The Grasses of Southern Queensland*. (Univ. of Qld Press, St Lucia, 1983.) Keys, descriptions, line drawings.

215 Trehane, P., *et al.* (eds), *International Code of Nomenclature for Cultivated Plants—1995*. (Quarterjack Publishing, Wimbourne, UK, 1995.)

216 Trigg, C., and Trigg, M., *Wildflowers of the Brisbane Ranges*. (CSIRO Publishing, 2000.) Mostly good colour photographs of almost 400 spp.

217 Urban, A., *Wildflowers and Plants of Central Australia*. (Portside Editions, Fishermans Bend, Vic., 1993.) Mostly two good colour photographs per page with brief descriptions arranged in families in systematic order following that of *Flora of Central Australia*. Covers the southern half of NT.

218 Usher, G., *A Dictionary of Botany*. (Van Nostrand, Melbourne, 1970.)

219 Walsh, N. G., and Entwisle, T. J., (eds), *Flora of Victoria*. Vols 2–4 (Inkata Press, 1994–99) Comprehensive and authoritative work covers all then-known species, with keys, descriptions, small distribution maps, numerous line drawings. Volume 1, edited by Foreman, D. B., and Walsh, N. G., (1993), contains introductory essays on Victorian climate, soils, weeds, rare and threatened flora, etc.

220 Walters, S. M., *et al.* (eds), *The European Garden Flora*. Vols 1–6. (Cambridge Univ. Press, Cambridge, UK, 1984–2000). A comprehensive and authoritative large-format work with keys and descriptions for all vascular plants cultivated in Europe. Vols 3–6 edited by Cullen, J., *et al.* Vol 2 the first published, vol. 1 reprinted 1990.

221 Webb, C., *et al.*, *Flowering Plants of New Zealand*. (DSIR Botany, Christchurch, 1990.) Introductory sections followed by discussions of 52 important plant families (each occupying two pages) accompanied by excellent colour photographs.

222 Wheeler, D. J. B., Jacobs, S. W. L., and Norton, B. E., *Grasses of New South Wales*. (Univ. of New England, Armidale, 1982.) Keys to the level of species, descriptions of genera, many line drawings.

223 Wheeler, J., *Wildflowers of the South Coast*. (Dept Conservation and Land Management, Como, 1996.)

Bibliography

224 Wheeler, J., *et al.*, *Flora of the South West*. (Univ. of WA Press, Nedlands, 2002.) Comprehensive work of two volumes, with descriptions, keys, and numerous small sketches of the plants in the Bunbury, Augusta, Denmark area. Includes notes on ecology and distribution.

225 Wheeler, J. R., (ed.), *et al.*, *Flora of the Kimberley Region* (Dept Conservation and Land Management, Como, WA, 1992.) Comprehensive and authoritative work with keys and descriptions for more than 2000 known species from the region. Numerous line drawings.

226 Whibley, D. J. E. and Symon, D. E., *Acacias of South Australia*. (Govt Printer, Adelaide, revised 2nd edn, 1992.) Comprehensive, very well illustrated with line drawings and colour photographs, detailed descriptions and keys. Originally publ. with Whibley as sole author.

227 White, M. D., *Coastal Vegetation of Anglesea–Airey's Inlet Region*. (Author publ., Anglesea, 1990.) Includes sketches and brief descriptions of more than 140 species arranged in groups, e.g., by flower colour and shape, to aid identification.

228 Wightman, G., and Andrews, M., *Plants of Northern Territory Monsoon Vine Forests*. (Conservation Commission of the NT, Darwin, vol. 1, 1989.) Good line drawings and brief descriptions of more than 70 species arranged to facilitate identification. Introductory sections include good colour photographs.

229 Wilding, J. L., *et al.*, *Crop Weeds*. (R. G., & F. J. Richardson, Meredith, updated edn, 1998.) Covers 117 weeds with brief descriptions, more than 340 colour and 100 black-and-white photographs illustrating various stages of growth. See also no. 163.

230 Williams, J. B., and Harden, G. J., *A Field Guide to the Rainforest Climbing Plants of New South Wales Using Vegetative Characters*. (Univ. of New England, Armidale, 1984.)

231 Williams, J. B., *et al.*, *Trees and Shrubs in Rainforests of New South Wales and Southern Queensland*. (Univ. of New England, Armidale, 1984.) Useful identification guide with brief descriptions and sketches of leaves and fruits, following a brief introduction.

232 Williams, K. A. W., *Native Plants of Queensland*. Vols 1–3. (Author publ., North Ipswich, 1979, Vols 2–3, 1987.) Each volume contains about 150 pages of good colour photographs, usually six per page, with brief descriptive captions on opposite page.

233 Willis, J. H., *A Handbook to Plants in Victoria*. Vol. 1, 2nd edn, vol. 2. (Melbourne Uni. Press, Carlton, 1970, 1972.) Detailed keys to all Victorian species then known, distributions, and references to illustrations in other books.

234 Willis, J. H., Fuhrer, B. A., and Rotherham, E. R., *Field Guide to the Flowers and Plants of Victoria*. (Reed, Sydney, 1975.) Over 400 colour photographs and brief text. An abridged version of Cochrane *et al.* (1973) (no. 50).

235 Wood, B., and Wood, D., *Flowers of the South Coast and Ranges of New South Wales. A Field Guide*. Vols 1–2. (Woods Books, Westangera, ACT, 1998, 1999.) Each volume with 400 good colour photographs, four per page, arranged by flower colour.

236 Woolcock, D., *A Fieldguide to the Native Peaflowers of Victoria and South-eastern Australia*. (Kangaroo Press, Kenthurst, and SGAP-NSW, 1989.) A small-format guide with descriptions and illustrations of 152 species, as well as 46 small colour

Bibliography

photographs. Includes an introduction to the structure of pea flowers, and summary of important features.
237 Woolcock, D. T., and Woolcock, C. E., *Australian Terrestrial Orchids*. (Nelson, Melbourne, 1984.) Includes an introduction to orchid structure, brief descriptions and many colour plates.
238 Wrigley, J. W., 'Australian Pea-Flowers', *Australian Plants*, 10: 95–123 (1979).
239 Wrigley, J. W., *Banksias, Waratahs, and Grevilleas, and all Other Plants in the Australian Proteaceae Family*. (Collins, Sydney, 1989.) Comprehensive, with introduction and cultivation notes, followed by brief descriptions and notes for each species, with small distribution map. Numerous line drawings and more than 45 pages of colour photographs
240 Wrigley, J. W., and Fagg, M., *Bottlebrushes, Paperbarks and Tea Trees, and all other Plants in the Leptospermum Alliance*. (Angus and Robertson, Pymble, 1993.) Introductory sections on the family, classification etc., illustrated with good colour photographs and some line drawings, with brief descriptions, distribution maps, ecological and cultivation notes.
241 Young, J., *Hakeas of Western Australia*. Vols 1–3. (Author publ., 1997, Vols 2–3, 2000.) Covering a total of nearly 170 species, each volume deals with different botanical regions. Species are described, and most illustrated with a colour photograph. No keys.

CD-ROMs

CD1 Backhouse, G., and Jeanes, J., *Wild Orchids of Victoria, Australia*. (Zoonetics, Seaford, 2000.) A pictorial guide with over 1300 good colour photographs of all then-recognised species (and some undescribed) grouped to facilitate comparison and identification. Includes brief descriptions, and notes on habitat and flowering times.
CD2 Brooker, M. I. H., *et al.*, *Euclid. Eucalypts of Southern Australia*. (CSIRO Publishing, Collingwood, 2nd edn, 2002.) An interactive key and information CD for all 690 species and subspecies currently known from the region. Includes descriptions, distribution maps, detailed notes and over 4000 colour photographs. Comprehensive and authoritative. First published as *Euclid: Eucalypts of South-eastern Australia*, 1997.
CD3 Hyland, B. P. M., *et al.*, *Australian Tropical Rainforest Trees and Shrubs. An Interactive Identification System for Trees and Shrubs*. (CSIRO Publishing, Collingwood, 1999.) Covers 1733 plants of northern Australian rainforests, extensively illustrated, with descriptions, and a computer key. See also previous edn listed as 122.
CD4 Maslin, B. R., (ed.), *Wattle. Interactive Identification of Australian Acacias*. (Dept Conservation and Land Management, and Aust. Biological Resources Study. Available from CSIRO Publishing, Collingwood, 2001.) Comprehensive electronic key to all known species, with detailed descriptions, numerous colour photographs, etc.

Bibliography

CD5 Sharp, D., and Simon, B. K., *AusGrass. Grasses of Australia*. (Australian Biological Resources Study, Canberra, 2002.) Available from CSIRO Publishing, Collingwood. An interactive identification and information guide to 1323 species of native and naturalised grasses. More than 4000 photographs and illustrations, 1400 maps, descriptions, and keys.

CD6 Thiele, K. R., and Adams, L. G., *The Families of Flowering Plants of Australia. An Interactive Identification Guide*. (Aust. Biological Resources Study, Canberra, 1999. Available from CSIRO Publishing, Collingwood.) A powerful electronic key in LucID format. Most characters are illustrated with small line sketches, and explained in accompanying notes. Descriptions of families, and colour photographs of representative species.

CD7 Viridans Pty Ltd, *Victorian Flora Database*. (Viridans Biological Databases, Brighton East, Vic., 1996.) Lists 4500 species of Victorian vascular plants with distribution data in grid map form which can be plotted on 11 map overlays. At least one colour photograph of each of 1100 species, and descriptive text for about 600.

Websites

This is but a very few of the websites related to plants. Numerous enthusiasts worldwide, often growers, have sites devoted to particular plant groups. These often include photographs of species and cultivars for sale or exchange. It is perhaps unwise to immediately accept as accurate the names of all images on the internet without checking against an authoritative text. Many botanic gardens and herbaria have sites, often making specialist information readily available. A recent development of note is Australia's Virtual Herbarium, an online botanical resource which is a collaborative project of the state, commonwealth and territory herbaria, and is accessed vai their websites. To date, label data from more than 40% of the herbarium specimens housed in these institutions have been databased. This will eventually involve over six million specimens. In time, the label data will be accompanied by, or have links to, species descriptions, illustrations, and photographs (of which some are already available). At present the main benefits are the ability to map the distributions of Australian species, and to check the currency of names.

Some sites require users to register before full access is granted. This might involve a cost.

W1 <home.vicnet.net.au/~sgapvic/> The homepage of the Australian Plants Society (Victoria) includes access to the journal *Australian Plants* online.

W2 <www.anbg.gov.au> The website of the Australian National Botanic Gardens in Canberra includes links to other sites providing technical botanical information (e.g., the International Code of Botanical Nomenclature) as well as a number of useful searchable databases including:
What's its name—a concise listing of Australian plant names and recent changes.

Bibliography

Australian Plant Name Index (APNI)—names and publication references for Australian higher plants.
Australian Cultivar Names—many names linked to colour photographs.
Australian Common Name Database

W3 <www.anbg.gov.au/abrs/flora/webpubl/glossaries.htm> Allows access to the glossary from the *Flora of Australia*. Many terms have associated illustrations or cross-references.

W4 <www.bgbm.org> The website of the Botanic Garden and Botanical Museum, Berlin, allows access to the latest version of the International Code of Botanical Nomenclature (the St Louis Code, 2000) at <www.bgbm.org/iapt/nomenclature/code/saintlouis/0000St.Luistitle.htm>

W5 <www.calm.wa.gov.au> Follow links to the Science Matters page and then to Florabase. This is a large database of names, distributions, photographs etc. of the WA flora.

W6 <www.ipni.org> The International Plant Names Index is a database of the names and associated basic bibliographic details of all seed plants.

W7 <www.nybg.org> The website of the New York Botanic Garden. Includes contact details for 2010 herbaria worldwide at <www.nybg.org/bsci/ih/ih.html>.

W8 <www.rbge.org.uk> The website of the Royal Botanic Gardens, Edinburgh. Allows access to the *Flora Europea* database at <www.rbg.org.uk/forms/fe.html>.

W9 <www.rbgkew.org.uk> The website of the Royal Botanic Gardens, Kew, England. Allows users to search the 'Kew Record of Taxonomic Literature', a large, invaluable, up-to-date database of taxonomic literature about plants, including much that can assist with identification. As most of these papers are published in scientific journals, users will need access to an academic library such as those attached to universities.

W10 <www.rbgsyd.gov.au> The website of the Royal Botanic Gardens, Sydney. Includes searchable plant databases.

W11 <www.rbg.vic.gov.au> The website of the Royal Botanic Gardens, Melbourne. Includes the *Census of Victorian Vascular Plants*, searchable for introduced species, and names added since the publication of the *Flora of Victoria*, among other categories. Also available is the index to *Muelleria*, the Gardens' journal, including abstracts to many recent articles.

Glossary

Some of the words listed have more than one meaning in biology but only the botanical usage has been included here. Many terms are used as nouns as well as adjectives, and both forms are listed if commonly used. The adjective is often derived from the noun by the addition of 'ate'.

abaxial	the surface of an organ (often of a leaf) facing away from the axis to which it is joined.
abort	failure to develop, e.g., of ovules that fail to develop into seeds.
abscission	the shedding of leaves, foliage branches, fruits or bark by the laying down of a special layer of cells.
achene	a dry, indehiscent, one-seeded fruit, produced from the superior ovary of a single carpel, but sometimes also applied to one-seeded fruits from an inferior ovary, as in the daisy family.
acicular	needle-shaped.
actinomorphic	of a flower from above, radially symmetrical (Fig. 2; Pl. 1a–c).
acuminate	narrowing to a point (Fig. 22).
acute	coming to a sharp abrupt point, e.g., as a leaf tip (Fig. 22), or of a style-arm in some daisies.
adaxial	the surface of an organ (often of a leaf) facing the axis to which it is joined.
adherent	with unlike parts or organs joined but only superficially, and without tissue continuity.
adnate	with unlike or like parts or organs integrally fused to one another with continuity of tissues.
adventitious	of tissues or organs developing in an abnormal position, e.g., adventitious roots arising from stems (Fig. 105).
aestivation	the arrangement of flower parts in the bud, particularly of the sepals and petals.
alien	a plant introduced from another region.
alternate	of leaves that arise singly at the nodes (Fig. 20).
amplexicaul	stem-clasping, usually of leaves at their point of attachment.
androecium	the stamens of a flower collectively (Fig. 1).

264

Glossary

androgynous	of a plant having male and female flowers in the same inflorescence. See Euphorbiaceae (Fig. 69c).
Angiosperms	a subgroup of the plant kingdom—the so-called flowering plants.
annual	a plant that completes its life cycle in one season.
anther	the pollen-bearing part of a stamen (Fig. 3a).
anthesis	the opening of a flower bud.
apetalous	without petals.
aphyllous	without leaves.
apical	pertaining to the apex or tip.
apocarpous	of a gynoecium made up of separate or free carpels (Figs 1, 4b–c).
appendage	an external outgrowth, often having no obvious function.
appressed	closely flattened to.
arborescent	tree-like in growth or general appearance.
aril	a fleshy, often coloured, outgrowth from a seed (adj. arillate).
aristate	bearing an awn (Fig. 22).
articulate	in general, jointed. Of a leaf, having a distinct point of attachment to the stem from which it arises (Fig. 20), cf. decurrent.
ascending	of leaves or branches curving upwards.
asepalous	without sepals.
asexual	of reproduction that does not involve gametes (q.v.)—that is, vegetative reproduction.
attenuate	narrowed, as in a leaf base (Fig. 22).
auricles	ear-shaped appendages, often found on leaves (Fig. 19, adj. auriculate).
awn	a needle-like appendage or stiff bristle (Figs 45, 79, 114; Pl. 11c).
axil	the angle between the upper surface of a leaf or bract and the stem to which it is attached (Fig. 18).
axile	attached to the central axis.
axillary	of buds or flowers arising in an axil (Fig. 18).
barbellate	minutely barbed.
bark	the protective external layer of tissue on the stems and roots of shrubs and trees.
basal	situated or attached at the base.
basifixed	of an anther, attached by its base to the filament (Fig. 3b).
beak	a long, pointed appendage (Fig. 109).
berry	a fleshy fruit, often containing many seeds, e.g., tomato.
biennial	a plant that develops vegetatively in the first year and produces a food-storage organ, then flowers and dies in the second year, e.g., carrots, onions.
bifid	forked into two parts.
bilabiate	having two lips, e.g., of a calyx or corolla (Figs 2f, 98).
bipinnate	of a compound leaf, with leaflets pinnately divided (Fig. 19).
bipinnatisect	a pinnatisect leaf with deeply dissected segments (Fig. 23).
bisexual	bearing both male and female sex organs.

Glossary

blade	the lamina or flattened part of a leaf (Fig. 19).
bract	a modified, often reduced, leaf usually associated with inflorescences or buds (Fig. 15, adj. bracteate).
bracteole	a small bract, often found on the pedicel or calyx (Figs 55, 61, 99).
bristle	a stiff hair.
bristle-gland	an elongated protruding gland with a multicellular wall. Possession of bristle-glands is an important feature distinguishing *Angophora* and *Corymbia* from other eucalypts.
bud	an undeveloped shoot, or inflorescence, or flower that will later expand and mature.
bulb	a short underground stem surrounded by fleshy leaf bases that store food material, and enclose one or more buds for the next season's growth.
caducous	non-persistent, often of floral parts, falling early.
caespitose	matted or tufted in growth habit.
calli (pl.)	of orchids, elongated outgrowths usually found on the labellum (Fig. 128b).
callus (sing.)	of some grasses, the hardened, usually pointed base of the lemma where it joins the rachilla.
calyptra	a cap.
calyx (pl. calyces)	the outermost whorl of the flower parts; the sepals collectively (Fig. 1).
calyx tube	a tube formed by fusion of the sepals, or sometimes meaning the floral tube.
campanulate	bell-shaped, often of a corolla (Fig. 2g).
capitate	having a head; head- or knob-shaped, e.g., of a stigma.
capitulum	a head of flowers, e.g., the inflorescence in most daisies (Figs 15, 101, 103).
capsule	a dry, dehiscent fruit developed from a syncarpous ovary.
carpel	the unit of the gynoecium made up of ovary, style and stigma (Figs 1, 4a).
caruncle	an outgrowth on a seed.
caryopsis	a dry one-seeded fruit with the pericarp fused to the testa, typical of the grasses.
catkin	an inflorescence type, a pendulous spike usually with small, bracteate, unisexual flowers, as seen in willows and birches.
caudicle	a stalk attaching the orchid pollen mass to the rostellum.
cauline	of leaves, growing from the stem (Fig. 20) (cf. radical).
cell	the structural unit of organisms. Also used as a synonym for loculus (q.v.).
chlorophyll	a green pigment in chloroplasts, essential for the process of photosynthesis.
chloroplast	an organelle, present in plant cells, that contains chlorophyll (see Chapter 5).

ciliate	having fine hairs (cilia) (Fig. 22).
circumcissile	of a fruit that breaks open around its circumference.
cladode	a flattened, photosynthetic stem.
clavate	club-shaped.
claw	the narrowed base of a petal or sepal (Fig. 2i).
coalescent	partially fused, more or less irregularly.
coccus (pl. cocci)	a one-carpel unit of a schizocarp (q.v.).
coherent	of like parts or organs joined, but only superficially.
column	the fused style, stigma and stamens, as in orchid flowers (Figs 126, 130). The basal, twisted part of an awn, as in some grasses (Pl. 11c).
compound head	a type of inflorescence in which small heads, each usually with an involucre, are sessile on a short axis, e.g., some members of the daisy family (Fig. 111; Pl. 10i).
compound leaf	a leaf having the blade divided into two or more leaflets (Fig. 19).
compound umbel	an inflorescence; literally, an umbel of umbels. A series of umbels whose stalks arise from one point (Fig. 12).
compression	of a grass spikelet, flattened either laterally or dorsally (Fig. 113).
connate	with like parts or organs joined by growth, e.g., united petals (Fig. 2g, h, j, k).
connective	the tissue connecting the two anther lobes (Fig. 3a).
connivent	parts converging or coming into contact, not fused.
contiguous	touching but not united.
convergent	of parts that come together gradually.
cordate	heart-shaped (Figs 21, 22).
coriaceous	leathery.
corm	a solid, reduced, underground stem of one year's duration, with thin leaf bases attached to the surface. The next season's corm forms on top of the older one (Fig. 121).
corolla	collectively, the petals of a flower (Fig. 1).
corolla tube	refers to the usually tubular basal portion of united petals (Figs. 2j, k).
corona	a trumpet-like outgrowth of the perianth, as in daffodils.
corymb	a flat-topped racemose inflorescence with the lower flower stalks longer than the upper, so all the flowers are more or less level.
corymbose	of inflorescences, resembling a corymb, i.e., flat-topped (Fig. 15g).
cotyledons	the seed-leaves attached to the embryo within the seed, sometimes containing stored food.
crenate	edged with rounded teeth (Fig. 23).
crenulate	finely crenate (Fig. 23).
culm	the inflorescence stalk of grasses and sedges (Fig. 116).
cultivar	difficult to define succinctly but usually thought of as a plant in cultivation whose origin or selection is primarily due to human intervention, and whose distinctive characteristics can be reliably reproduced. Usually given a cultivar name which is written in single quotes (see Chapter 6).

Glossary

cuneate	wedge-shaped (Figs 21, 22).
cyathium	a reduced inflorescence, surrounded by an involucre, and resembling a single flower, e.g., Euphorbia (Fig. 69; Pl. 6a).
cyme	one of a group of inflorescence types in which there is a terminal flower, then further flowers on new lateral shoots (Fig. 16, adj. cymose).
cypsela	a small, indehiscent, one-seeded fruit, derived from an inferior ovary, typical of daisies (Figs 106, 109).
deciduous	seasonal shedding of leaves, or of small appendages such as hairs or stipules shed before the organ on which they are growing.
decurrent	of a leaf, with the base extending down the stem (Fig. 20).
decussate	of opposite leaves, with each successive pair at right angles to the pair below (Fig. 20).
dehiscence	the splitting, or opening of an anther or fruit to release the pollen or seeds respectively (adj. dehiscent).
deltate	triangular (Fig. 21).
deltoid	3-dimensional figure with triangular faces, but often used as synonymous with deltate in referring to a two-dimensional triangular shape.
dentate	with a toothed margin (Fig. 23).
denticulate	with a finely toothed margin (Fig. 23).
dichasium	a cymose inflorescence in which an axis ends in a flower and two lateral buds develop behind it. A dichasium is said to be compound if this pattern is repeated (Fig. 16b).
dichotomous	of branching, forked, in which the branches are of equal size and often branch again in the same way.
dicots	abbreviation of dicotyledons.
dicotyledons	in classification, the subgroup of flowering plants having seeds with two cotyledons. (Formally spelled Dicotyledones. See Chapter 6.)
digitate	diverging from a single point, like the fingers of a hand (Fig. 115).
dioecious	with the male and female sex organs on separate plants.
diploid	of an organism, having two complete sets of chromosomes, one from each parent (2n). Cf. haploid (n).
disc	in a flower, a structure usually associated with the ovary either at the top (Fig. 12) or base (Figs 7, 62), or lining the floral tube, and often secreting nectar. In a radiate head of a daisy, the tubular florets collectively.
disc florets	the tubular florets in the centre of a radiate head of a daisy.
dissected	of a leaf, with the lamina divided into deep lobes (Fig. 23).
distichous	of leaves, arranged in two opposite rows (Fig. 20).
divergent	of anthers that spread apart at the base (Fig. 3d).
dorsal	at the back.

dorsifixed	of an anther, with the filament attached to the back of the anther (Fig. 3c; Pl. 1c).
double	of flowers, having more than the usual number of petals; common in horticultural varieties.
drupe	a fleshy fruit with a hard endocarp enclosing the seed, a fleshy mesocarp and the exocarp soft but forming a 'skin', e.g., cherry or plum.
e-	prefix meaning without.
elliptical	ellipse-shaped; e.g., of a leaf where length is about two to three times width, broadest in the centre and tapering to each end (Fig. 21).
emarginate	lacking a distinct margin, or with a notch at the tip (Fig. 22).
embryo	the rudimentary plant within a seed.
endemic	of a plant, naturally restricted to a single area or country.
endo-	a Greek prefix meaning within.
endocarp	the innermost wall of the pericarp, which often becomes hard and stony as in plums and cherries.
endosperm	a food reserve, associated with the embryo in the seeds of flowering plants (adj. endospermic).
entire	of a margin that is smooth and undivided (Fig. 23).
ephemeral	a plant that completes its life cycle in a short time when conditions are favourable.
epi-	a Greek prefix meaning upon.
epicalyx	a whorl of sepal-like bracts outside and often attached to the calyx, the bracts usually alternate with the sepals (Fig. 11).
epicormic	of buds that lie dormant under the bark of some trees, such as eucalypts, and grow when the tree is defoliated.
epicotyl	the part of the plant axis or stem between the cotyledonary node and first foliage leaves (Fig. 18).
epidermis	the outermost layer of cells, usually single, on leaves, young shoots etc.
epigynous	of a flower when the floral tube completely encloses the ovary, and fuses with the ovary wall so that the other flower parts arise above the ovary (Figs 8b, 12).
epipetalous	of stamens that are attached to the petals (Figs 88–92, 102d).
epiphyte	a plant growing on another but not deriving food from it.
epitepalous	of stamens that are attached to the tepals as in grevilleas (Fig. 33; Pl. 2a) and some monocots (Fig. 123).
ericoid	of leaves that are small, narrow, rigid and often pungent, as in the heath family.
exocarp	the outer layer of the pericarp, often the skin of fleshy fruits.
exotic	an introduced plant, not native. See also alien.
exserted	protruding, e.g., of anthers protruding from a corolla.
exstipulate	without stipules.

Glossary

extra-axillary	of a bud that is not produced in a leaf axil.
falcate	sickle-shaped (Fig. 21).
family	a taxonomic group of one or more genera, a subdivision of an order. The names of most botanical families end in '-aceae'.
filament	the stalk of an anther (Fig. 3a).
filiform	thread-like. A type of tubular floret in some daisies.
fimbriate	with a fringed margin.
flora	collectively, the plants of an area, or a book about the plants of an area, usually with descriptive entries following a system of classification.
floral leaves	the bracts or bracteoles associated with inflorescences.
floral tube	in a flower, a structure appearing as the united bases of perianth parts and stamens, and often tubular. Also referred to as the receptacle, receptacle tube, thalamus, thalamus tube, torus, hypanthium, and calyx tube (Figs. 9, 14, 47).
floret	a small flower, mainly used in referring to the flowers of the daisy and grass families.
follicle	a dry fruit, containing one to many seeds, derived from a flower with a single carpel and splitting along one side at maturity. The pericarp is hard and tough, or leathery. Fruit type of *Banksia* and *Grevillea* and some other members of the family Proteaceae.
free	not joined laterally to a similar member, e.g., free petals.
free central	of placentation, the placenta arising centrally from the base of a unilocular ovary (Fig. 5h).
fruit	the ripe ovary of a flowering plant, sometimes including accessory structures (e.g., the beak and pappus of dandelion, Fig. 109).
funicle	the stalk of an ovule attaching it to the placenta, and later the stalk of the seed.
gamete	a haploid (q.v.) cell involved in sexual reproduction.
gamopetalous	of a flower, with united petals.
gamosepalous	of a flower, with united sepals.
garden escape	a plant introduced for gardens, which now grows and reproduces freely outside cultivation. See naturalised and alien.
geniculate	bent like a knee.
generic name	the name of a genus, e.g., *Acacia*, *Eucalyptus*.
genus	in classification, the main rank between family and species. A taxonomic group of species which resemble each other more closely than they resemble other groups.
glabrous	smooth, without hairs or other surface covering.
gland	an organ or tissue, that secretes oil, nectar, resin or water.
glaucous	of leaves, stems etc., that have a bluish appearance, usually caused by a waxy substance on the surface.
glume	each of the two bracts at the base of a grass spikelet (Figs 114, 116), one is present in sedges.

gynandrous	of a flower in which the stamens and styles have united to form a column as in Orchidaceae (Fig. 130c, d).
gynobasic	of a style that arises from near the base of a deeply lobed ovary (Figs 6c, 98).
gynoecium	the female part of a flower, consisting of one or more free or united carpels (Figs 1, 4).
gynophore	a stalk which raises the gynoecium above the points of insertion of the sepals, petals and stamens (Figs 36b, 44b).
habit	the external form and shape of a plant.
haploid	having one set of chromosomes (n) cf. diploid (2n).
hastate	halberd-shaped (Fig. 21).
haustorium	in parasitic plants, a structure developed for penetrating the host's tissues (Pl. 4f).
head	an inflorescence of small, usually sessile flowers, usually surrounded by an involucre (Figs 15, 101, 103).
herb	a non-woody plant but not a moss or liverwort. A plant of medicinal or culinary value that is not necessarily non-woody.
herbaceous	soft and green, non-woody.
herbarium	a collection of preserved, usually dried, plant material. Also a building in which such collections are stored.
hermaphrodite	having functional male and female organs in the same flower or on the same plant.
hesperidium	a type of berry, an orange, lemon, etc.; the flesh is made up of swollen fluid-containing hairs, which fill the loculi.
hilum	the scar on the testa at the point where the seed broke away from its stalk (Fig. 18).
hyaline	thin and translucent.
hybrid	usually the offspring of two different varieties or species.
hypanthium	the floral tube. See also receptacle, torus.
hypocotyl	the part of the plant axis below the colytedonary node but above the root (Fig. 18).
hypogynous	of a flower in which the other parts arise around the base of the ovary (Figs 1, 8a). Of an organ, arising at the base of the ovary, e.g., a gland (Figs 33, 35).
imbricate	of organs that overlap, e.g., bracts or petals.
imparipinnate	of a pinnate leaf with an uneven number of pinnae including one terminal pinna (Fig. 19h).
incised	with sharply pointed lobes, e.g., of a leaf (Fig. 23).
incurved	curved upwards, e.g., of a leaf margin (Fig. 22).
indehiscent	of fruits that do not split open to release the seeds.
indigenous	native to an area.
indumentum	collectively the hairs or scales (etc.) on the surface of plant organs.
induplicate	folded inwards.

Glossary

indusium	a cup-like structure enclosing the stigma, e.g., in the family Goodeniaceae (Fig. 99; Pl. 3c).
inferior	of an ovary, with perianth parts and stamens attached above it. Epigynous flowers (Figs 8b, 12) have inferior ovaries.
inflorescence	a flowering shoot of more than one flower.
internode	the section of stem between two successive nodes (Fig. 18).
invested	of a surface covered with hairs or scales.
involucral bract	a member of an involucre.
involucre	ring(s) of bracts, usually around an inflorescence, typical of daisies (Figs 101, 103).
involute	having inrolled margins, e.g., of a leaf (Fig. 22).
irregular	of a flower, asymmetrical (Figs 2d, e, f).
juvenile leaf	a seedling leaf that differs markedly from the adult leaf (Fig. 82).
keel	of the corolla in pea flowers, the two $\pm$ united lower petals (Fig. 55). In general, a ridge along a fold, e.g., at the back of a grass lemma (Fig. 114e).
labellum	a petal of an orchid flower, usually the lowest, which is different in form from the two lateral ones (Figs 126, 127). Sometimes used to describe modified petals in other groups also.
labiate	of a corolla, with one or more petals forming a lip.
lamina	the blade or flat part of a leaf (Fig. 19), or bract.
lanceolate	a shape, e.g., of a leaf—where length is about five times width, the widest part below the centre, and tapering to each end (Fig. 21).
latex	a milky juice that exudes from some plants such as milk thistles and dandelions.
leaf	an outgrowth of a stem, usually flat and green, the main function being food manufacture by photosynthesis.
leaflet	a segment of a compound leaf (Fig. 19).
legume	a dry fruit derived from a single carpel, which splits along both sutures at maturity (Figs 53, 57).
lemma	the larger and outermost of the two bracts of a grass floret (Fig. 114; Pl. 11c).
liane	a woody climbing or twining plant.
lignotuber	a woody rootstock produced below ground level by some plants.
ligule	of grasses, a membranous outgrowth or ring of hairs at the junction of the leaf sheath and leaf blade (Fig. 114). Of the corolla of some florets in daisies, an elongated strap-shaped extension (Fig. 109). Adj. ligulate.
limb	the wide upper part of a petal; or the upper, usually spreading, section of a corolla of united petals (Fig. 2h, i).
linear	long and narrow (Fig. 21).
lobe	part of a dissected leaf. The free upper part of organs fused at the base, e.g., calyx lobe, corolla lobe (Figs 2g, 23).
loculus	a chamber or cavity, e.g., within an ovary (Fig. 4a).

Glossary

lodicule	in grasses, one of the small scales at the base of the ovary, often interpreted as a reduced perianth member (Fig. 117).
lomentum	a pod-like indehiscent fruit that develops constrictions between the seeds and at maturity breaks into one-seeded segments.
lyrate	lyre-shaped. Of a pinnatifid leaf, having a large, rounded terminal lobe and successively smaller lateral lobes (Fig. 23).
marginal	of a vein, close to the edge of a leaf. Of placentation, the ovules attached along one side of the ovary (Fig. 5a).
mealy	of a surface, with a whitish floury appearance (Pl. 6d–e).
membranous	thin and flexible, not green and herbaceous.
mericarp	a one-carpel unit of a fruit which breaks up at maturity
meristem	a group of actively dividing cells.
mesocarp	the middle layer of a pericarp (q.v.).
midrib	the large central vein of a leaf (Fig. 19).
monochasium	a cymose inflorescence in which the axis ends in a flower and further growth arises from a bud behind the terminal flower. A monochasium is said to be compound if the pattern is repeated (Fig. 16a).
monocots	abbreviation of monocotyledons.
monocotyledons	in classification, the subgroup of the flowering plants having seeds with a single cotyledon. (Formally spelled Monocotyledones. See Chapter 6.)
monoecious	having male and female flowers on the same plant.
monopodial	of branching, when a main axis continues to grow, producing lateral buds which become flowers or shoots that may repeat the same pattern, as in a racemose inflorescence (Fig. 15). Cf. sympodial.
morphology	the shape or form of an organism.
mucro	a short, sharp point, usually on a leaf, often fragile (Fig. 22).
mucronate	of a leaf, with a short, sharp point (Fig. 22).
naturalised	describing a plant, introduced from another region, that grows and reproduces readily in competition with the natural flora.
nectar	a sugary liquid often produced by flowers, or occasionally by glands elsewhere on the plant.
nectary	a gland that produces nectar (Figs 7, 14).
nerve	a vein, e.g., in a leaf or bract.
node	the part of a stem from which leaves emerge (Fig. 18).
non-endospermic	lacking an endosperm.
nut	a hard, dry, indehiscent, one-seeded fruit derived from a syncarpous ovary, such as an acorn.
nutlets	one-seeded entities produced by a fruit that fragments at maturity.
obcordate	of a leaf, heart-shaped with the widest part towards the tip (Fig. 21).
oblanceolate	a shape, e.g., of a leaf—of length about five times the width, the widest part above the centre and tapering to the base (Fig. 21).

Glossary

oblique	of a leaf blade, asymmetrical at the base (Fig. 22).
oblong	a shape, e.g., of a leaf—where length is two or three times width and sides roughly parallel (Fig. 21).
obovate	a shape, e.g., of a leaf—where length is about one and a half times width and widest above the centre (Fig. 21).
obtuse	rounded or blunt (Fig. 22).
ochrea	a sheath around a stem formed from united stipules (Fig. 20).
operculum	a lid—see section on Eucalyptus (Figs 80, 83).
opposite	of leaves, in pairs at each node (Fig. 20). Of organs in a flower (e.g., stamens and sepals, Fig. 1), when on the same radius.
orbicular	a shape, e.g., of a leaf—$\pm$ circular (Fig. 21).
oval	a shape, e.g., of a leaf—about twice as long as wide, rounded at both ends (Fig. 21).
ovary	the hollow portion at the base of a carpel containing one or more ovules (Fig. 4a).
ovate	a shape, e.g., of a leaf—of length about one-and-a-half times width, widest below the centre. Egg-shaped in outline (Fig. 21).
ovules	structures in the ovary, which enclose the egg-cells and become seeds after fertilisation.
palea	the smaller and inner bract of a grass floret (Fig. 117; Pl. 11c).
palmate	of a leaf, with 5–8 leaflets spreading from the same point, like the fingers of a hand (Fig. 19).
palmatifid	of a deeply dissected leaf in which the lobes spread like the fingers of a hand (Fig. 23).
panicle	a branched raceme, with each branch being a raceme. In general, a much-branched inflorescence (Fig. 15c, adj. paniculate).
papillae	small, rounded projections (adj. papillose).
pappus	the hairs, bristles or scales (usually forming a ring) found at the base of the corolla in florets of most daisies and sometimes interpreted as a modified calyx (Figs 102, 103, 108, etc.).
parasite	a plant living on another plant, the host, and drawing food from it.
parietal	of placentas, attached to the wall of the loculus (Fig. 5g).
paripinnate	of a pinnate leaf having an even number of pinnae including a pair in the terminal position (Fig. 19i).
partite	divided into parts, e.g., a five-partite corolla has five petals.
pedicel	the stalk of an individual flower (Fig. 1, adj. pedicellate).
peduncle	the stalk of an inflorescence (Figs 15, 83, adj. pedunculate).
peltate	of a leaf or other organ that has the stalk attached to the middle of the lower surface (Figs 22, 66).
pendulous	of an ovule, attached at or near the top of the ovary (Fig. 5b). Of an anther when hanging from the filament (Fig. 3d).
pepo	a type of berry formed from an inferior ovary and containing many seeds, usually large with a tough outer skin, e.g., pumpkin, cucumber.

Glossary

perennating	of an organ that survives vegetatively from season to season. A period of reduced activity between seasons is usual.
perennial	a plant with a normal life span of more than two years.
perfoliate	of a leaf base that completely surrounds the stem, so the stem appears to pass through the leaf (Fig. 20).
peri-	a Greek prefix meaning around.
perianth	the outer, non-reproductive part of the flower; usually consisting of a whorl of sepals and/or a whorl of petals, or two whorls of tepals (Fig. 1).
pericarp	the fruit wall, developed from the ovary wall.
perigynous	of a flower, with a cup-shaped floral tube with the perianth parts and stamens attached to the rim of the cup (Figs 8c, 47).
petal	a member of the inner whorl of perianth parts (Fig. 1).
petaloid	like a petal, usually colourful and showy.
petiole	a leaf stalk (Fig. 20, adj. petiolate).
phloem	the part of the vascular tissue that conducts synthesised nutrients. See Chapter 5.
photosynthesis	the process by which sugars are made from carbon dioxide and water in cells containing chloroplasts; the chemical energy required is produced from solar energy in the presence of the pigment chlorophyll. See Chapter 5.
phyllary	a bract in an involucre, as in daisies.
phyllode	a leaf-like organ derived from the petiole; the foliage of many wattles is of phyllodes (Figs 48–51).
pinna (pl. pinnae)	a leaflet of a pinnate leaf, or the equivalent part of a bipinnate leaf (Figs 19, 52).
pinnate	of a compound leaf, having a row of leaflets on each side of the rachis (Fig. 19).
pinnatifid	a shape, e.g., of a leaf—when pinnately lobed with the blade cut about half way to the midrib (Fig. 23).
pinnatisect	a shape, e.g., of a leaf—when pinnately lobed, the blade cut almost to the midrib (Fig. 23).
pinnule	a segment formed by the pinnate division of a pinna (Figs 19, 52).
pistil	the gynoecium, or a single carpel of an apocarpous gynoecium.
placenta	the tissue within the ovary to which the ovules are attached (Fig. 5).
placentation	the arrangement of placentas within the ovary (Fig. 5).
plicate	pleated
plume	feather-like (adj. plumose).
plumule	the rudimentary shoot of an embryo.
pollen	the collective name for the pollen grains that develop within the anther. The grains contain the male reproductive nuclei.
pollination	the process of pollen transfer from the anther to the stigma.
pollinium	a mass of pollen enclosed by a membrane, or held together by a sticky substance. During pollination it is transferred as a unit (Fig. 127).

Glossary

polygamous	having bisexual and unisexual flowers on the same or different plants.
pome	a fruit that has developed partly from the ovary wall but mostly from the floral tube, e.g., apple.
prickle	a hard, pointed appendage without vascular tissue.
procumbent	of a plant, with stems that lie along the ground.
prop roots	adventitious roots that develop from the lower nodes of a stem and help to support it.
prostrate	procumbent (q.v.).
protandrous	of a flower that sheds its pollen before the stigma is receptive.
protogynous	of a flower in which the stigma is receptive before the anthers of that flower release their pollen.
pseudobulb	usually of orchids, a part of the stem (usually basal) when thickened or swollen.
pubescent	covered with fine hairs.
pulvinus	a swelling at the base of a petiole or leaflet concerned with movement in response to stimuli such as excessive wind (Figs 51, 52).
pungent	having a sharp, hard point.
raceme	an inflorescence of stalked flowers with the youngest at the top (Fig. 15, adj. racemose).
rachilla	the axis of a divided pinna of a bipinnate leaf (Fig. 19), or the axis of a grass spikelet (Fig. 114).
rachis	the axis of a pinnate leaf to which the leaflets are attached (Fig. 19), or the main axis of an inflorescence.
radiate	of a capitulum, with ray florets surrounding disc florets (Fig. 101; Pl. 10a–e).
radical	of leaves, arising from near the base of the plant (Fig. 20).
radicle	the rudimentary root of an embryo within a seed.
rambler	a plant, lacking erect stems, which spreads along the ground or over other plants, fences etc.
ray floret	one of the ligulate florets around the edge of a capitulum, as in many daisies (Fig. 101).
receptacle	the end of the flower stalk to which the flower parts are joined, considered by some authors to include the floral tube, if present (Fig. 1).
receptacle tube	see floral tube.
recurved	curved under, e.g., of a leaf margin (Fig. 22).
regular	of flowers that are radially symmetrical or actinomorphic (Fig. 2a–c).
reniform	kidney-shaped (Fig. 21).
repent	of a prostrate plant that roots at the nodes of the spreading stems. Creeping.
retuse	with a central notch in a bluntly rounded apex (Fig. 22).
revolute	rolled under, e.g., of leaf margins (Fig. 22).

reticulate	net-like.
rhizome	a perennial underground stem, which is usually horizontal.
rhomboidal	a shape, e.g., of a leaf—when roughly diamond-shaped and length = width (Fig. 21).
rim	of the fruit of eucalypts, the scar (usually enlarged) remaining after the operculum is shed at the opening of the flower (Figs. 80b, 83c).
root cap	a cap of cells that protects the end of a root as it grows through the soil.
root hairs	outgrowths of the outermost layer of cells just behind the root tip, functioning as water-absorbing organs.
rosette	leaves radiating from a centre, usually basal as in the dandelion (Fig. 109).
rostellum	an outgrowth from the column of an orchid flower (Fig. 127).
rosulate	with a basal rosette.
rotate	of a corolla, with a short tube and spreading limb; wheel-shaped (Fig. 2h).
rudimentary	of an organ that is incompletely developed.
rugose	of a surface, wrinkled and usually rough.
runcinate	of a lobed leaf, with the ends of the lobes pointing backwards (Fig. 23).
sagittate	shaped like an arrow-head with basal lobes pointing downwards (Fig. 21).
salver-shaped	of a tubular corolla, with lobes that spread more or less horizontally, as in *Phlox*.
samara	a dry, indehiscent one-seeded fruit with part of the wall extended to form a wing, e.g., elm and ash fruits.
saprophyte	a plant living on, and absorbing its food from, dead organic matter and usually not carrying out photosynthesis.
scabrous	of a surface, covered by small, rough projections.
scale	often used instead of bract for small, flat, leaf-like appendages.
scape	the flowering stem of an otherwise stemless plant, e.g., dandelion (Figs 109, 122).
scarious	thin and dry, often used to describe such margins of otherwise herbaceous bracts.
schizocarp	a dry fruit derived from a syncarpous ovary and splitting at maturity into one-seeded segments.
scorpioid cyme	a cyme that branches so that the inflorescence becomes curved in one direction like a scorpion's tail.
secund	with the lateral members all turned to one side, often describing one-sided inflorescences.
seed	the reproductive unit of a plant, the product of a fertilised ovule and containing an embryo with food reserves.
seed-leaf	a cotyledon, a leaf-like organ within the seed.

Glossary

seedling	the young plant that develops after germination of a seed.
self-compatible	self-fertile, that is, a flower's own pollen will fertilise its ovules.
self-incompatible	self-infertile, that is, a flower's own pollen will not fertilise its ovules.
self-pollination	also called selfing, the acceptance by stigmas of pollen from the same flower or from flowers on the same plant, which means they are self-compatible.
semi-inferior	of an ovary which extends above and below the point of attachment of the perianth parts and stamens (Fig. 73d, Pl. 3c).
sepal	a unit of the calyx, or outer perianth whorl, commonly green in colour (Fig. 1).
sepaline	resembling a sepal, of the sepals.
sepaloid	resembling a sepal.
septicidal	of dehiscence in which a fruit splits along lines coinciding with the septa.
septum (pl. septa)	the wall(s) dividing an ovary into two or more loculi (Fig. 4).
serrate	of a margin, regularly toothed (Fig. 23).
serrulate	finely serrate, with small teeth (Fig. 23).
sessile	without a stalk, most often of leaves or flowers (Fig. 20).
sheath	a leaf base that forms a tubular casing around the stem (Figs 19, 114).
shoot	usually the aerial part of a plant. A stem including its dependent parts, leaves, flowers etc.
shrub	a perennial woody plant, with no single axis, which branches freely and does not exceed 10 m (approx.) in height.
silicula	a dry fruit derived from a bilocular ovary that splits from the bottom up, with the seeds left attached to the septum. The length of the fruit is less than three times its width.
siliqua	as for the silicula, except that the length is more than three times the width.
simple	of leaves, undivided, but the margin may be dissected (Figs 19, 23).
sinuate	margins shallowly and smoothly indented (Fig. 23).
solitary	single, of flowers that grow one per plant, one in each axil, or widely separated on the plant.
spadix (pl. spadices)	an inflorescence composed of a spike of flowers on a fleshy axis and subtended by a large bract called a spathe, e.g., arum lily (Pl. 6b).
spathe	a large bract enclosing a spadix (Pl. 6b). Also used with reference to bracts subtending the inflorescence in some monocotyledon families (Figs 122, 124).
spathulate	spoon-shaped (Fig. 21).
species	a kind of organism, that is, a group of individuals that is distinguishable from all other closely related groups. The basic unit of classification (see Chapter 6).

specific epithet	the second word in the two-part scientific name of an organism; cf. trivial name.
spicate	spike-like.
spike	an inflorescence of sessile flowers on an unbranched axis (Fig. 15a).
spikelet	the unit of the inflorescence of grasses (Fig. 114). Also present in sedges and some other monocot families.
spine	a modified branch or leaf or stipule that has become hard and pointed.
spur	a short shoot; or a conical or tubular outgrowth from the base of a perianth segment, often containing nectar.
stamen	the male reproductive organ of the flowering plants consisting of a stalk or filament and a bilobed anther (Fig. 3).
staminal ring	of the fruit of eucalypts, the tissue to which the stamens were attached in the flower (Fig. 84b).
staminate	of a flower, with fertile stamens but no functional carpels.
staminode	a sterile stamen, usually of modified structure.
standard	the large petal that stands at the back of the flower in the pea family (Fig. 58).
stellate	star-like, used to describe the tuft of hairs arising from a central point, as in *Pomaderris* (Fig. 73) and *Thomasia* (Fig. 74).
stem	the plant axis, either aerial or subterranean, which bears leaves, branches and flowers.
stigma	the receptive surface on the style on which pollen can germinate, often papillose and sticky.
stipe	a stalk (adj. stipitate).
stipel (pl. stipellae)	one of two small leaf-like appendages at the base of the leaflets of some compound leaves.
stipitate	with a stalk.
stipule	one of a pair of appendages found at the base of some petioles (Figs 20, 56, 60, 74; Pl. 7b, adj. stipulate).
stolon	a creeping stem that may bear a new plant at its tip, e.g., strawberry. Underground stolons often end in tubers (q.v.).
stoloniferous	having stolons.
stomata	the pores, in the outer cell layer of plants, that are involved in gas exchange, most prevalent on leaves.
style	the section of a carpel or syncarpous gynoecium between the ovary and the stigma (Figs 1, 4, 6). Not always developed, and the stigma is then borne on the top of the ovary (Pl. 3a).
style-arms (-branches)	the parts of a branched style (Figs. 69c, 70b, 102b, 108a).
subtend	to stand beneath or close to, as in a bract at the base of a flower.
subulate	awl-shaped (Fig. 21).
superior	of an ovary, with the perianth parts and stamens, or their united bases, arising below it. Hypogynous (Fig. 8a) or perigynous flowers (Fig. 8c) have superior ovaries.

Glossary

suture	a line along which fruit may split open.
sympodial	of branching, when a main axis stops growing, subsequent growth arising from the development of lateral buds just behind the apex as in cymose inflorescences (Fig 16). Cf. monopodial.
syncarpous	of a gynoecium, made up of united carpels.
syngenesious	of stamens that are united by their anthers to form a tube around the style (Figs 3k, 102d).
tailed	of anthers, having a long slender basal appendage (Fig. 108c).
tap root system	a root system with a prominent main root and smaller lateral or branch roots (Fig. 109).
taxon (pl. taxa)	a general term for any category in classification; thus species, genera and families are all taxa.
taxonomy	the science of the classification of organisms.
tendril	a leaf or stem modified to coil around outside supports, enabling the plant to climb (Fig. 56).
tepal	a perianth segment not recognised as a petal or sepal.
terete	cylindrical (Fig. 22).
terminal	situated at the tip.
testa	the protective coat of the seed.
thalamus	the receptacle of a flower, sometimes a synonym for floral tube.
thalamus tube	see floral tube.
thallus	a simple plant body not differentiated into a root-shoot system. A term more often encountered in connection with non-flowering plants.
throat	the opening at the top of a tubular corolla or perianth (Fig. 2j–k).
tomentose	covered with dense matted hairs.
toothed	of a leaf, deeply and sharply lobed (Fig. 23).
torus	the receptacle of a flower; sometimes a synonym for floral tube.
trichome	a general term for an epidermal outgrowth, a hair or scale etc.
trifoliolate	a compound leaf of three leaflets, e.g., a clover leaf (Fig. 19f).
tripinnate	of leaves which are three times pinnately divided (cf. bipinnate Fig. 19j).
trivial name	the second word in the two-part scientific name of an organism; cf. specific epithet.
truncate	with an abruptly blunt end (Fig. 22).
trunk	the upright, large main stem of a tree.
tuber	a swollen underground stem or root containing stored food, forming a seasonal perennating organ.
tubercle	a small, rounded protruberance.
type genus	in nomenclature, the genus from which the family name is derived.
umbel	an inflorescence in which the flower stalks arise at one point (Figs 15f, 80, 122).

understorey	a general term for the plants growing under the canopy of taller ones.
undulate	usually of a leaf, with a wavy margin (Fig. 22).
unilocular	having one loculus, e.g, the ovary in the families Proteaceae, Fabaceae (Figs 5a–c, 33, 36, 58, etc.).
unisexual	of a flower, having reproductive parts of only one sex (Figs 38, 70, 121). Of an inflorescence or plant, having flowers of one sex.
urceolate	urn-shaped; that is, like a rounded vase swollen in the middle.
valvate	usually of petals or sepals, having margins adjacent but not overlapping.
valves	of dehiscent fruits, flaps of tissue which open to release the seeds (Figs. 80b, 82b, 83c).
veins	the vascular bundles as they are seen in leaves.
venation	the arrangement of veins, usually either parallel or reticulate.
versatile	of an anther, when its point of attachment to the filament is small, allowing easy movement (Fig. 3c).
verticillate	arranged in whorls, especially of leaves.
villous	having long, weak hairs.
whorl	several organs arising from the same level around an axis, such as a whorl of sepals or petals, or leaves (Figs 1, 20).
whorled	of leaves, three or more around the stem arising from one node (Fig. 20).
wings	of the corolla in pea flowers, the two lateral petals (Fig. 58). In general, flattened outgrowths of any plant part may be referred to as wings (Fig. 130c).
xylem	the water-conducting tissue of vascular plants. See Chapter 5.
zygomorphic	of flowers that are asymmetrical and can be divided into equal halves along one longitudinal plane only (Fig. 2d–f).
zygote	a fertilised egg-cell.

Index

Acacia, 110
  acinacea, Fig. 48
  botrycephala, see Fig. 52
  diffusa, see Pl. 7a
  floribunda, Fig. 49
  genistifolia, Pl. 7a
  iteaphylla, Fig. 50
  longifolia, Pl. 7c, 238–9
  melanoxylon, 110
  myrtifolia, Pl. 7b
  paradoxa, 111
    stipules, 44
  pycnantha, Fig. 51
  sophorae, see Pl. 7c, 238–9
  terminalis, Fig. 52
Acaena, 103
  anserinifolia, 238
  echinata, 103, Fig. 45
  fruit, 39
  novae-zelandiae, 103, 238
achene, 193, 195
  beaked, 195
Acmena, 153
Acrotriche serrulata, Pl. 2e
actinomorphy, 11
Adenanthos terminalis, Fig. 26
aestivation, 11
Agapanthus, 224

  praecox, Figs. 122–3
  stamens, 13
Agonis, 153
Albizia lophantha, 238
alders, pollination, 35
ALLIACEAE, 210
Allium, 210, 224
  triquetrum, 224
  vineale, 224
Allocasuarina, 72, 238
  verticillata, Pl. 4 c–d
  perianth, 11
  unisexual flowers, 23
allspice, 153
almond, 103
Alopecurus pratensis, Pl. 11d
AMARYLLIDACEAE, 210, 224
Amaryllis belladonna, 224
amaryllis family, 224
Amperea xiphoclada, 139
Amyema, 87
  pendulum, Pl. 4a–b
Anagallis, placentation, 17
androecium, 7, 11
Anemone, inflorescence, Fig. 16
angel's trumpet, Pl. 8g, 183
Angiospermae, 2
Angiosperms, 55

282

Angophora, 153
Anguillaria dioica, see Fig. 121
annual plants, 43
anthers, 7, 11, Fig. 3
    appendages, 11, 129
        in daisies, 195
    attachment to filament, 11
    dehiscence, 11
    tailed, 195
Anthophyta, 2, 55
apple-berry, 98, Fig. 42
apple box, 153
apples, 103–4
    fruit wall of, 38
Arachis hypogea, 120
Arbutus, 171
Artemisia dracunculus, 193
Arthropodium, 219
    strictum, 239, Fig. 118
artichoke, 193, Pl. 10j
articulation, 213, Fig. 113
ash, bud scales, 44
    fruit dispersal, 40
Asparagus, 219
    asparagoides, 219
    officinalis, 219
Asphodelus fistulosus, 219
Aster, 193
ASTERACEAE, 193, Pl. 10
    stamens, 13
Asterolasia, 129
Atriplex, 88
    cinerea, Pl. 5a
    nummularia, 88
    unisexual flowers, 23
authorities, for names, 55
Avena sterilis, Pl. 11a–c
avocado, 94
axil, of leaf, 44
axis, in plants, 42
Azalea, 171

baby's breath, inflorescence, Fig. 16
bacteria, nitrogen fixing, 120
bamboo, 211
Banks, Sir Joseph, 55
Banksia, 72
    marginata, Figs. 27, 28
    gynoecium, 15
banksia, silver, Figs. 27, 28
barley, 211
bean, 120
    broad, 120
    carob, 117
    cotyledons, 37
    french, 120
    germination, 39
    germination, early growth, Fig. 18
    seed structure, 39, Fig. 18
beard-heath, 171
    common, Fig. 91
    pink, Fig. 90
beard-orchid, purplish, Fig. 127
beauty-heads, lemon, Figs. 111–12
beetroot, 88
bell-climber, orange, 98, Fig. 41
Bentham, George, 55
Beta, 88
bidgee-widgee, 103
    fruit, 39
Billardiera, 98
    bignoniacea, 98, Fig. 41
    scandens, 98, Fig. 42
billy-buttons, golden, Pl. 10i
binomials, 54
birches, fruit dispersal, 40
    pollination, 35
bird-orchid, common, Fig. 128
blackberries, 103–4
blackwood, 110
bloodwood, 155
blue gum, classification of, 55
blue-lily, tufted, 219, Fig. 120

Index

blue pincushion, 191, Pl. 6c
blue squill, Fig. 119
blue stars, 219, Fig. 119
bluebell creeper, 98
blue-bush, 88
  pearl, Pl. 5f
  three-wing, Pl. 5e
boneseed, 193, Pl. 10b
Bontia, 188
boobialla, 188
  common, Pl. 9b
  creeping, 188
  sticky, Pl. 9a
bootlace bush, 151
Boronia, 129
  denticulata, Pl. 1c
  mollis, Fig. 62
  gynoecium, 15
boronia, soft, Fig. 62
bottlebrush, 153
  crimson, Figs. 77–8
boxthorn, 183
Brachychiton, 147
bracteoles, 44
bracts, 44
  involucral, 194–5
    Pimelea, 151
bramble, small-leaf, 103
branching, monopodial, 31
  sympodial, 31
briar, sweet, 103
Bromus catharticus, Fig. 114
  unioloides, see Fig. 114
broom, 120
broom rush, Pl. 12b–c
broom spurge, 139
Brown, Robert, 55
brown-beards, Fig. 127
Brugmansia, 183
  × *candida*, Pl. 8g
Brunonia australis, 191, Pl. 6c

BRUNONIACEAE, 191
brush box, 153
buckthorn, 144
  family, 144
  italian, 144
bulbs, 43
burrs, 103, 121
Bursaria spinosa, 98
bursaria, sweet, 98
bush-pea, 120
  golden, Figs. 60–1
  stipules, 44
buttercup, 92, Pl. 1f
  gynoecium, 15

cacti, stems, 43
Caesalpinia, 117
CAESALPINIACEAE, 109, 117
Caesalpinioideae, 109
Caladenia, 231
  dilatata, 239, see Fig. 126
  phaeoclavia, 239, Fig. 126
Caleana, 231, 239
  minor, Pl. 11g
Calectasia, 218
Calendula, 193
  officinalis, Pl. 10d
calla, Pl. 6b
Callistemon, 153
  citrinus, Figs. 77–8
Calocephalus citreus, Figs. 111–12
Calochilus robertsonii, Fig. 127
Calostemma, 224
Calytrix, 154
  tetragona, Fig. 79
calyx, 7, 9
  bilabiate, 11
  petaloid, 147
  tube, 21
campion, inflorescence, Fig. 16
candlebark, 157

cape gooseberry, calyx, 9
cape tulip, 228
cape wattle, 110
capeweed, 193
capitulum, of daisies, 194
Capsicum, 183
Cardwellia sublimis, 72
carob, 117
carpels, 7, 13
  number of, 17
  structure, Fig. 4
Carthamus, 193
cassava, 139
Cassia, 117
  artemisioides, 239, see Figs. 53–4
  fistula, 117
cassia, silver, Figs. 53–4
Cassytha, 94
  glabella, 94
  melantha, 94, Pl. 4f
castor oil plant, seeds, 37
Casuarina, 72
  stricta, 238, see Pl. 4c–d
CASUARINACEAE, 72
catchfly, inflorescence, Fig. 16
cat mint, 185
caudicle, 231
CD-Roms, 261
cells, 42
cellulose, 42
Ceratonia, 117
Cestrum, 183
  parqui, Pl. 8a
Chaenomeles, 103
  speciosa, Fig. 46
Chamaescilla, 219
  corymbosa, Fig. 119
Chamelaucium, 153
CHENOPODIACEAE, 88, Pls. 5, 6d–e
  seeds, 37
Chenopodium album, 88

cherry, flower type, 21
  fruit structure, 38
  Japanese flowering, Fig. 47
chillies, 183
Chiloglottis gunnii, 239, see Fig. 128
  valida, 239, Fig. 128
chlorophyll, 42
chloroplasts, 42, 50
Choisya, 129
Christmas tree, Western Australian, 87
chromosomes, 42
Chrysanthemoides monilifera, 193, Pl. 10b
Chrysanthemum, 193
Cichorieae, 196
Cichorium endiva, 193
Cinnamomum camphora, 94
  zeylanicum, 94
cinnamon, 94
  fungus, 51
citrus, 129
Citrus, 129, 239
  australasica, 129
  glauca, 129, 239
cladodes, 43
classification, 2–3, 52
claw, of petal, 9
Clematis, 92
  microphylla, 92, Fig. 38
  perianth, 9
  petioles, 43
  unisexual flowers, 23
clematis, small-leaved, 92
clover, 120
  white, Pl. 7e
cloves, 153
club-rush, knobby, Pl. 12a, d
cluster, of flowers, 34
coconut palm, fruit, 40
Coleonema, 129, 239
Coleus, 185

285

Index

column, 191, 230–1
Comesperma volubile, stems, 43
COMPOSITAE, 193, Pl. 10
compression, in grasses, 213, Fig. 113
computer keys, 68
conifers, pollination, 35
connective, 11
Conospermum, 73
 mitchellii, Figs. 29, 30
 stamens, 13
coral-pea, 120
corms, 43
corolla, 7, 9
 actinomorphic, 11
 bilabiate, 11
 clawed, 9
 colour of, 9
 irregular, 11
 limb of, 9
 lobes, winged, 189
 regular, 11
 throat of, 9
 tube, 9
 types, Fig. 2
 zygomorphic, 11
corona, 9
correa, common, Fig. 63
Correa, 129
 reflexa, Fig. 63
 corolla, 9
 fruit, 38
Corymbia, 155
 citriodora, 155
 ficifolia, 155
 gummifera, 155
 maculata, 155
Cotoneaster, 103
cotyledonary node, 39, Fig. 18
cotyledons, 37, 39, 53, Fig. 18
Craspedia chrysantha, 239, see Pl. 10i
Crassula, gynoecium, 15

 multicava, Pl. 1a
CRASSULACEAE, gynoecium, 15
Crataegus, 103–4
 monogyna, 103
 oxyacanthoides, 103
Crinum, 224
Crocus, 228
Crowea, 129
 stamen, Fig. 64
Cryptandra, 144
 amara, Fig. 72
 stamens, 13
cryptandra, 144
 bitter, Fig. 72
 spiny, Fig. 72
cucumbers, unisexual flowers, 23
cultivars, 56–7
Cunningham, Allan, 55–6
cyathium, 139
cymes, 31
Cynara scolymus, 193, Pl. 10j
Cynodon dactylon, Fig. 115
CYPERACEAE, 213, Pl. 12
Cyperus eragrostis, Pl. 12e–f
cypsela, 193, 195
Cytisus, 120
cytoplasm, 42

daffodil, 224
 corona of, 9
Dahlia, 193
daisies, Pl. 10
 family, 193
 fruit dispersal, 40
daisy-bush, musk, Fig. 103
daisy, Easter, 193
 Michaelmas, 193
dandelion, 196, Figs. 109–10
Daphne, 151
'Dargen Hill Monarch', Fig. 104
Datura, 183

Index

Daviesia, see Pl. 7e
desert lime, 129
Dichopogon strictus, 239, see Fig. 118
dicotyledons (dicots), 2, 53
Dillenia, 150
DILLENIACEAE, 150
Dillwynia, 120
  glaberrima, Fig. 55
  sericea, Pl. 7e
diosma, 129
Diosma, 239
Diplarrena, 228
discs, 18
dissection, 59–61
Diuris, 231
dodder-laurel, 94
  coarse, 94, Pl. 4f
  slender, 94
dormancy, 39
Dracophyllum, 171
Drapetes tasmanica, 151
Drosera, 95
  glanduligera, Pl. 4e
  leaves, 44
DROSERACEAE, 95
Duboisia hopwoodii, 183, Pl. 8h
duck-orchid, large, 231
  small, 231, Pl. 11g
dusty miller, 144
  stamens, 13

early nancy, 219, Fig. 121
  unisexual flowers, 23
elms, fruit dispersal, 40
embryo, 37, 40
  sac, 36
emu-bush, 188, Pl. 9c–f
Enchylaena, 88
  tomentosa, Pl. 5e
endive, 193
endosperm, 37, 39

EPACRIDACEAE, 171
  key to genera, 67
Epacris, corolla tube, 21
  floral formula, 24
  impressa, 171, Figs. 88–9, Pl. 2d
    corolla colour, 9
    stamens, 13
epicalyx, 9, Pl. 9g
epicormic shoots, 158
epicotyl, 39
Eremocitrus glauca, 129, 239
Eremophila glabra, Pl. 9c
  maculata, Pl. 9d
  ovata, Pl. 9e
  platycalyx, Pl. 9f
Erica, 171
ERICACEAE, 171
Eriostemon, 129
  myoporoides, 239, see Fig. 65
  key to species, 65
eucalypts, 154
  germination, 39
  classification, 155
Eucalyptus, 153–4
  camaldulensis, Figs. 80–1
  classification, 155
  globulus, Fig. 82
    classification, 55
  leucoxylon, Figs. 83–4
  leucoxylon 'Rosea', see Fig. 84
  obliqua, 155, 157
  pauciflora, 154
  regnans, 159
  rubida, 157
  viminalis, 157
  structure, 157
Eugenia, 153
Euphorbia, 139
  characias, 139, see Pl. 6a
  peplus, 139, Fig. 69
  pulcherrima, 139

Index

wulfenii, 139, Pl. 6a
EUPHORBIACEAE, 139
everlasting, 193, Fig. 104

Faba, 120
FABACEAE, 109, 120, Pl. 7
  gynoecium, 15
Fabales, 109
Faboideae, 109
families, in classification, 2, 54
  names, 56
    pronunciation, 3, 56
fan-flower, 189
  coast, Pl. 3c
fat hen, 88
Feijowa sellowiana, 153
fennel, Fig. 12
  perianth, 11
fertilization, 36, Fig. 17
Ficinia nodosa, Pl. 12a, d
fig, fruit structure, 38
filament, 7, 11
finger lime, 129
firewheel tree, 72
flag, butterfly, 228
flat-pea, common, Fig. 59
flat-sedge, drain, Pl. 12e–f
flatweed, Pl. 10h
floral formulae, 24–5
floral tube, 21, 103, Fig. 9
Floras, 52, 57
florets, of daisies, 193–4
  of grasses, 211–13
flower, arrangement of parts, 6, 18
  basic structure of, 7, Fig. 1
  epigynous, 21, 104
  hypogynous, 19, 104
  numbers of parts, 23
  papilionaceous, 121
  perigynous, 21, 103
  sections of, Fig. 25

types of, Fig. 8
flowering plants, 2
  classification, 52
flowers, arrangement of, 31
  solitary, 31
  unisexual, 23
Foeniculum vulgare, Fig. 12
  floral formula, 25
  perianth, 11
form, in classification, 54
fossils, 1, 53
foxtail, meadow, Pl. 11d
Fragaria, 103
frangipani, native, 98
Fraxinus, bud scales, 44
Freesia, 228
fringe-lily, 219
  common, Pl. 11e
fringe-myrtle, Fig. 79
fruits, common, 38, 103
  development, 37
  dispersal, 40
  in daisies, 195
  structure, 37, 103–4
  types, 37
Fuchsia magellanica, Fig. 14
  floral tube, 21
fuchsia, native, Fig. 63
funicle, 15
furze, stamens, 13

Galinsoga parviflora, Pl. 10a
gallant soldier, Pl. 10a
gametes, 35
  male, 36
garden plants, names of, 56
garland lily, 224
garlic, 210, 224
  three-cornered, 224
  wild, 224
Gaultheria, 171

geebung, prickly, Fig. 36
genera, in classification, 2, 54
generic characters, 54
  names, 54
genus, in classification, 2, 54
Gladiolus, 228
  corms, 43
gland flower, Fig. 26
glands, on bracts, 139
  on phyllodes, 111
glasswort, beaded, Fig. 37
Gleditsia, 117
glumes, of grasses, 213, Pl. 11a–b
Glycine max, 120
Goodenia, 189
  ovata, Fig. 99
goodenia, hop, Fig. 99
GOODENIACEAE, 189
  corolla, 9
gorse, stamens, 13
GRAMINEAE, 211, Pl. 11
grapefruit, 129
grape hyacinth, 219
grape, tendrils, 43
grass, couch, Fig. 115
  prairie, Fig. 114
grasses, 211
  fruit wall, 37
  pollination, 35
grass-flag, 228
  pretty, Fig. 125
grass-tree, austral, Pl. 12h–i
grass-trees, 210, 218
greenhood, 231
  dwarf, Fig. 131
  tall, Fig. 130
Grevillea, 72, Pl. 2b
  robusta, 72, Figs. 31–2
  rosmarinifolia, Fig. 33, Pl. 2a
  floral formula, 25
  gynoecium, 15

perianth, 9
stamens, 13
style-end, 18
grevillea, rosemary, Fig. 33
ground-cherry, sticky, Pl. 2c
groundsel, common, Figs. 105–6
  purple, Pl. 7c
guava, 153
guinea-flower, 150, Fig. 75, Pl. 1b
gum, large-fruited yellow, Figs. 83–4
  lemon-scented, 155
  river red, Figs. 80–1
  southern blue, Fig. 82
  spotted, 155
  Tasmanian blue, Fig. 82
  Western Australian flowering, 155
gum trees, see *Eucalyptus*, 154
gynoecium, apocarpous, 15
  structure, 7, 13
  syncarpous, 15
gynophore, 104
Gypsophila, inflorescence, Fig. 16

Hakea, 72
  decurrens, 238, Figs 34–5
  sericea, 238, see Figs. 34–5
  style-end, 18
Halosarcia pluriflora, Pl. 5d
hand lens, use of, Fig. 24
Hardenbergia, 120
haustoria, 87, 94
hawthorn, 103
head, heterogamous, 194
  homogamous, 194
  of daisies, 193–4
heath, 171
  common, 171, Figs. 88–9, Pl. 2d
  corolla colour, 9
  stamens, 13
  family, 171
  peach, Fig. 92

Index

swamp, 172, Figs. 94–5
heather, 171
hedge wattle, 111
　stipules, 44
Helianthus, 193
Helichrysum, 193
　bracteatum, 239, see Fig. 104
henbit dead-nettle, 185
Hevea brasiliensis, 139
Hibbertia, 150
　australis, see Fig. 75
　empetrifolia, Pl. 1b
　riparia, 239, Fig. 75
　stricta, 239
Hibiscus, epicalyx, 9
　rosa-sinensis, Pl. 9g
　stamens, 13
hilum, 37, Fig. 18
Homeria, 228
honey locust, 117
honey-myrtle, 153
　giant, Pl. 6f
honey-pots, Pl. 2e
horehound, 185
hyacinth, 219
Hyacinthus, 219
hybrids, names of, 57
Hymenosporum flavum, 98
hypanthium, 21
Hypericum leschenaultii, Pl. 1e
Hypochoeris radicata, Pl. 10h
hypocotyl, 39

indusium, 189, 191, Fig. 99, Pl. 3c
inflorescences, 31
　cymose, Fig. 16
　racemose, Fig. 15
insectivorous plants, 95
internodes, 43, Fig. 18
involucre, of daisies, 194
IRIDACEAE, 228

Iris, 228
iris, family, 228
　white, 228
Isolepis nodosa, see Pl. 12a, d
Isopogon, 73
Ixia, 228

japonica, Fig. 46
jonquil, 224
　corona of, 9
Juncus sarophorus, Pl. 12b–c

kangaroo apple, 183, Pl. 8c
keel (petals), 121
Kennedia, 120
key routes, *Acaena echinata*, 104
　Epacridaceae, 172
　Proteaceae, 74
　Rutaceae, 130
　Salicornia, see *Sarcocornia*, 89
　Sarcocornia quinqueflora, 89
　Thomasia petalocalyx, 147
keys, using, 61
　computer, 68
key to genera, Epacridaceae, 67
　Rutaceae, 64
key to species, *Eriostemon*, 65
Kunzea ambigua, Fig. 85
kunzea, white, Fig. 85
kurrajong, 147

labellum, 191, 230–1
LABIATAE, 185
Lachenalia, 219
Lactuca sativa, 193
lamb's ears, 185
LAMIACEAE, 185
　corolla, 11
Lamium, 185
　amplexicaule, 185
Lasiopetalum, 147

latex, 196, Pl. 6a
Lathyrus, floral formula, 25
  odoratus, 120, Figs. 56–8
  stipules, 44
LAURACEAE, 94
laurel, bay, 94
  camphor, 94
  family, 94
Laurus, 94
  nobilis, 94
Lavandula dentata, 185, 239
  stoechas, 185
lavender, 185
  Spanish, 185
  topped, 185
leaf primordia, 51
leaves, 43, Figs. 19–23
  adult/juvenile, 157
  apices, Fig. 22
  arrangement, 44, Fig. 20
  attachment, Fig. 20
  bases, Fig. 22
  compound, 44, Fig. 19
  ericoid, 172
  margins, Figs. 22, 23
  scale-like, 72
  shapes, Fig. 21
  simple, 44, Fig 19
  venation, 44
Lechenaultia, 189
leek-orchid, 231
leeks, 224
legumes, 109, 121, Pl. 7
Leguminales, 109
LEGUMINOSAE, 109
lemma, 212
lemons, 129
Leonotis, 185
Leptorhynchos squamatus, Figs. 107–8
Leptospermoideae, 153
Leptospermum, 153

floral tube, 21
  myrsinoides, Fig. 86
  ovary, 21
  scoparium, Pl. 3d
lettuce, 193, 196
Leucodendron, 72
Leucojum, 224
  aestivum, Fig. 124, Pl. 3b
Leucopogon, 171
  ericoides, Fig. 90
  virgatus, Fig. 91
Leucospermum, 72
Libertia, 228
  pulchella, Fig. 125
lignotuber, 158
LILIACEAE, 210, 219
Lilium, 219
lilly-pilly, 153
lily, 219
  arum, 44, Pl. 6b
  belladonna, 224
  chocolate, 219, Fig. 118
  family, 219
  garland, 224
  Murray, 224
lily of the valley, pampas, 183, Pl. 8f
limes, 129
Linaria, spur, 9
Linnaeus, 2, 54
lion's ear, 185
Lissanthe strigosa, Fig. 92
loculus, 13, 15
lodicules, 211–12
loganberry, 103
Lolium perenne, Figs. 116–17
Lomandra, 210, 218
lomentum, 121
longitudinal section, 61
Lophostemon, 153, 154
LORANTHACEAE, 87
Loranthus, 87

Index

love-creeper, stems, 43
lucerne, 120
Lychnis, inflorescence, Fig. 16
Lycium ferocissimum, 183
Lycopersicon esculentum, 183, Pl. 8e
Lyperanthus nigricans, 239, see Fig. 129

Macadamia, 72
magnification, of illustrations, 5
Magnolia soulangeana, Fig. 10
  flower parts, 7
Magnoliaceae, gynoecium, 15
Magnoliophyta, 2, 55
Magnoliopsida, 55
Maireana, 88
  sedifolia, Pl. 5f
  triptera, Pl. 5c
maize, 211
mallee roots, 158
mallees, 153, 158
mallow, wax, Fig. 11
Malococera, 88
Malus, 103
Malvaceae, epicalyx, 9, Pl. 9g
Malvaviscus arboreus, Fig. 11
Manihot esculenta, 139
manna gum, 157
manuka, Pl. 3d
Marianthus bignoniaceus, see Fig. 41
marigold, Figs. 101–2
  English, 193
  garden, Pl. 10d
marjoram, 185
marrows, unisexual flowers, 23
Marrubium vulgare, 185
mat-rushes, 210, 218
medic, 120
  fruit, 40
Medicago, 120
  fruit, 40
Melaleuca, 153
  armillaris, Pl. 6f

linariifolia, Fig. 87
  stamens, 13
Mentha, 185
meristems, 51
Mesembryanthemum, leaves, 44
messmate, 155, 157
Mexican orange, 129
Microcitrus, 129, 239
midge-orchid, 231
Mimosa, 110
MIMOSACEAE, 109, 110, Pl. 7a–c
  gynoecium, 15
Mimosoideae, 109
mint, 185
  corolla, 11
  family, 185
mint-bush, 185
  round-leaf, Fig. 97
mistletoe bird, 87
mistletoe, drooping, Pl. 4a–b
  European, 87
mistletoes, 87
monocotyledons (monocots), 2, 53, 210
mountain ash, 159
Muscari, 219
MYOPORACEAE, 188
Myoporum, 188
  floribundum, 188
  insulare, 188, Pl. 9b
  parvifolium, 188
  platycarpum, 188
  sp., Pl. 9a
myoporums, 188
MYRTACEAE, 153
myrtle, 153
  family, 153
Myrtoideae, 153
Myrtus, 153

names, changes of, 56
  meanings of, 55
  pronunciation of, 3, 56

Index

naming system, binomial, 54
Narcissus, 224
  corona of, 9
nectaries, extra-floral, 111
nectary, 18
nectary gland, 17
needlewood, bushy, Figs. 34, 35
Nepeta, 185
Nerine, 224
Nicotiana, 183
  alata, Pl. 2f
  suaveolens, Pl. 8i
  tabacum, 183
nightshade, black, 183, Pl. 8b
nightshades, 183
nodes, 43
Nomenclature, International Code of Botanical, 56
nucleus, 42
Nuytsia, 87

oaks, pollination, 35
oat, sterile, Pl. 11a–c
ochrea, Fig. 20
oil, aromatic, 157, 185
  castor, 139
oil glands, in leaves, 129, 157
Olearia argophylla, Fig. 103
onion grass, 228
onions, 43, 210, 224
onion-weed, 219
operculum, in eucalypts, 156
  in *Richea*, 171
oranges, 129
orchid, family, 230
  red-beak, Fig. 129
ORCHIDACEAE, 230
orchids, 230, Figs. 126–31, Pl. 11g–i
orders, in classification, 2, 54, 56
  names of, 56
    pronunciation of, 3, 56
Origanum, 185

ovary, 7
  inferior, 21, Pl. 3b
  position of, 21, Fig. 8
  semi-inferior, 21, Pl. 3d
  structure, 13, Fig. 4
  superior, 19, Pls. 1a–f, 2a
  T.S., Fig. 25, Pl. 2f
ovule, 7, 15
  pendulous, 17
  structure, 36

painted nettle, 185
palea, 212
panicle, Fig. 15
  spike-like, 34, Pl. 11d
Papaver
  nudicaule, Pl. 3a
  placentation, 17
  sepals, 9
paper-bark, 153
paper-flower, 147, Fig. 74
PAPILIONACEAE, 109, 120
  gynoecium, 15
Papilionatae, 109
Papilionoideae, 109
pappus, 193, 195
Paracaleana, 239
Paraserianthes lophantha, 110, 238
parasites, 87, 94
parrot-pea, 120
  showy, Pl. 7e
  smooth, Fig. 55
Passiflora, gynoecium, 15
  placentation, 17
passion flower, gynoecium, 15
  placentation, 17
Patersonia, 228
pea, cotyledons, 37
  family, 120
    gynoecium, 15
  garden, 120
  germination, 39

293

Index

seed structure, 39
sweet, 120, Figs. 56–8
peanut, 120
pears, 103, 104
pedicel, 7
peppers, 183
perianth, 7, 9
 fruiting, 38, 88
perianth parts, 9
perianth tube, 23
Persea americana, 94
Persoonia, 73
 juniperina, Fig. 36
petals, 7, 9
 gland-like, 147
 hooded, 144
 imbricate, 11
 valvate, 11
Petunia, 183
Phaseolus vulgaris, 120
phebalium, forest, Fig. 66
Phebalium squamulosum, Fig. 66
Philotheca myoporoides, 63, 129, 239, Fig. 65
phloem, 50
photosynthesis, 50
phyllodes, 110
Physalis, 183
 calyx, 9
 viscosa, Pl. 2c
Pimelea, 151
 axiflora, 151
 glauca, Fig. 76
 floral tube, 21
Pimenta dioica, 153
pimpernel, placentation, 17
pincushion, blue, Pl. 6c
pineapple guava, 153
pineapple, fruit structure, 37
pink-bells, 97, Fig. 40, Pl. 1d
pistil, 15

Pisum sativum, 120
pitcher plants, leaves, 44
PITTOSPORACEAE, 98
Pittosporum angustifolium, 98, 238, Fig. 43
 phillyreoides, 238
 undulatum, 98, Fig. 44
 calyx, 9
pittosporum, family, 98
 sweet, 98, Fig. 44
 calyx, 9
 weeping, 98, Fig. 43
pituri, 183, Pl. 8h
placentation, 17, Fig. 5
Platylobium obtusangulum, Fig. 59
plum, 103
Poa, 211
POACEAE, 211
Podolepis jaceoides, Pl. 11e
podolepis, showy, Pl. 11e
poinsettia, 139
poison-berry, green, Pl. 8a
pollen, 7, 11, 36
 compatibility of, 37
 germination of, 36
pollen-tube, 36
pollination, 35, Fig. 17
pollinators, 35
pollinia, 231
Pomaderris, 144
 oraria, 239
 paniculosa, 239, Fig. 73
 ovary, 21
pomaderris, coast, Fig. 73
pome, 104
poppy, Iceland, Pl. 3a
 placentation, 17
 sepals, 9
potato, 183
potato bush, velvet, Pl. 8d
potato vine, 183

Index

potato weed, Pl. 8d
Prasophyllum, 231
  spicatum, Pl. 11h
prickles, 104
primrose, placentation, 17
Primula, placentation, 17
PRIMULACEAE, placentation, 17
Prionotes, 171
Prostanthera, 185
  rotundifolia, Fig. 97
Protea, 72
  as type genus, 56
protea family, 72
PROTEACEAE, 72
  gynoecium, 15, Pl. 2a
Prunus, 103
  flower type, 21
  serrulata, Fig. 47
pseudobulb, 230
Psidium guajava, 153
Pterostylis, 231
  longifolia, 239, see Fig. 130
  melogramma, 239, Fig. 130
  nana, Fig. 131
Pultenaea, 120, see Pl. 7e
  gunnii, Figs. 60–61
  stipules, 44
pulvinus, 111
pumpkins, unisexual flowers, 23
purple-flag, 228
Pycnosorus chrysanthes, 239, Pl. 10i
Pyrorchis nigricans, 239, Fig. 129
Pyrus, 103

quince, 103, 104
  flowering, Fig. 46

radicle, 39
ragwort, 193
RANUNCULACEAE, 92
  gynoecium, 15

Ranunculus, 92, Pl.1f
raspberries, 103, 104
raspberry, native, 103
receptacle, 7
  of daisies, 193
receptacle tube, 21
reproductive organs, 11
Rhagodia, 88
  baccata, see Pl. 6d
  candolleana, Pl. 6d
  parabolica, Pl. 6e
RHAMNACEAE, 144
  stamens, 13
Rhamnus, 144
  alaternus, 144
  catharticus, 144
rhizomes, 43
Rhododendron, 171
  laetum, Pl. 2g
rice, 211
rice-flower, 151
  smooth, Fig. 76
Richea, 171
  pandanifolia, 172
  procera, Fig. 93
Ricinocarpos pinifolius, 139, Fig. 70
  floral formula, 25
Ricinus communis, 139
Romnaldia, 218
Romulea rosea, 228
root cap, 51
root hairs, 51
roots, 50
Rosa, 103, 104
  rubiginosa, 103
ROSACEAE, 103
  flower type, 21
rose, 103
  family, 103
  flower type, 21
  hips, fruit wall of, 38

Index

stipules, 44
rosemary, 185
rose-of-China, Pl. 9g
Rosmarinus, 185
rostellum, 231, Pl. 11i
Rubus, 103, 104
   fruticosus, 103
   parvifolius, 103
rue, 129
   family, 129
rush, broom, Pl. 12b–c
Ruta, 129
   graveolens, 129
RUTACEAE, 129
   gynoecium, 15
   key to genera, 64
rye-grass, perennial, Figs. 116–17

sage, 185
   wild, Fig. 98
Salicornia quinqueflora, 89
Salpichroa origanifolia, 18, 183, Pl. 8f
saltbush, coast, Pl.5a
   fragrant, Pl. 6e
   old man, 88
   ruby, Pl. 5e
   seaberry, Pl. 6d
saltbushes, 88, Pl. 5
   seeds, 37
Salvia, 185
   verbenaca, Fig. 98
samphires, 88
sandalwood, false, 188
sandalwoods, 87
SANTALACEAE, 87
Sarcocornia quinqueflora, 89, Fig. 37
Scaevola, 189
   pallida, Pl. 3c
   spinescens, 189
scaly buttons, Figs. 107–8
Sclerolaena, 88

SCROPHULARIACEAE, corolla, 11
seablite, austral, Pl. 5b
sections, of flowers, 4, 61, Fig. 25
sedge, umbrella, Pl. 12e–f
sedges, 213, Pl. 12
Sedum, floral formula, 24
   gynoecium, 15
   spectabile, Fig. 13
seed coat, 37
seed-leaves, 37, 53
seeds, development of, 37
   dispersal, 40
   dormancy, 39
   germination, 39–40
   stratification, 39
   structure, 39
Senecio elegans, Pl. 10c
   vulgaris, Figs. 105–6
Senna, 117
   artemisioides, Figs. 53–4
senna, 117
sepals, 7
septum, 15
she-oak, drooping, Pl. 4c–d
she-oaks, 72
   perianth, 11
   pollination, 35
sheep's burr, 103, Fig. 45
   fruit, 40
shoots, epicormic, 158
Silene, inflorescence, Fig. 16
silky oak, 72, Figs. 31–2
smoke-bush, 73
   stamens, 13
   Victorian, Figs. 29, 30
snapdragons, corolla, 11
snowflake, 224, Fig. 124, Pl. 3b
snow gum, 154
snow-in-summer, Fig. 87
SOLANACEAE, 183
Solanum aviculare, 183

ellipticum, Pl. 8d
jasminoides, 183
laciniatum, 183, Pl. 8c
nigrum, 183, Pl. 8b
stamens, 13
tuberosum, 183
Sollya heterophylla, 98
Sonchus oleraceus, Pl. 10f–g
soybean, 120
Sparaxis, 228
spathes, 44
species, in classification, 2, 54
specific epithet, 54
spider-orchid, 231
  green-comb, Fig. 126
spikelets, 213, Pls. 11b, 12d, f, g
spinach, 88
Spinacia, 88
Spiraea, 103
spotting characters, 70
Sprengelia, 171
  incarnata, Figs. 94–5
  corolla, 9
spur, 9
spur velleia, 189
spurge, Pl. 6a
  broom, 139
  petty, 139, Fig. 69
spurges, 139
Spyridium, 144
  parvifolium, stamens, 13
St. John's wort, Pl. 1e
Stachys, 185
Stackhousia, 142
  monogyna, Fig. 71
  corolla, 9
stackhousia, creamy, Fig. 71
STACKHOUSIACEAE, 142
stamens, 7, Fig. 3
  epipetalous, 13
  epitepalous, 13

structure, 11, Fig. 3
unilateral, 228
union of, 13
staminodes, 13
standard (petal), 121, Figs. 57–8
stems, modified, 43
  types, 43
Stenocarpus, 72
Sterculia, 147
STERCULIACEAE, 147
stigma, 7, 13, 18, Fig. 4
stinkwood, Figs. 67–8
stipules, 44, Figs. 19–20
  as spines, 111
  leaf-like, 147
stolons, 43
stonecrop, Fig. 13
  gynoecium, 15
stratification, 39
strawberries, 103, 104
style, 7, 13, 18, Fig. 4
  gynobasic, 18
  insertion on the ovary, Fig. 6
style-arms, 18
  in daisies, 195
style-end, 18
STYLIDIACEAE, 191
Stylidium, 191
  graminifolium, Pl. 9h, 239
  sp. 2, 239, Fig. 100
Stypandra caespitosa, 239, see Fig. 120
Suaeda australis, Pl. 5b
subspecies, in classification, 54
sugar beet, 88
sundews, 95
  leaves, 44
sundew, scarlet, Pl. 4e
sun-orchid, 231, Pl. 11i
swamp-heath, 171, Figs. 94–5
sweet pea, 120, Figs. 56–8
  stipules, 44

Index

tendrils, 44
sword-sedge, coast, Pl. 12a, g
symmetry, of flowers, 11
Syzygium aromaticum, 153

Tagetes 'Cinnabar', Figs. 101–2
tamarind, 117
Tamarindus, 117
tapioca, 139
tap root, 50
Taraxacum officinale, Figs. 109–10
tar bush, Pl. 9c
tarragon, 193
tea-tree, 153, Pl. 3d
 silky, Fig. 86
Tecticornia, 88
 verrucosa, Pl. 5g
Telopea, 72
tendrils, 43, 121
tepals, 9, 210
testa, 37
Tetratheca, 96
 ciliata, 97, Figs. 39–40, Pl. 1d
thalamus tube, 21
Thelionema, 219
 caespitosum, 239, Fig. 120
Thelymitra, 231
 aristata, Pl. 11i
 grandiflora, see Pl. 11i
thistle, milk, Pl. 10f–g
 sow, Pl. 10f–g
thistles, 193, 195, 196
 fruit dispersal, 40
Thomasia, 147
 petalocalyx, 147, Fig. 74
thornapple, 183
thorns, 104
Thryptomene, 153
thyme, 185
Thymelaea, 151
THYMELAEACEAE, 151
Thymus, 185

Thysanotus, 219
 tuberosus, Pl. 11e
tinsel-lily, 218
toadflax, spur, 9
tobacco, 183
 austral, Pl. 8i
 T.S. ovary, Pl. 2f
tomato, 183, Pl. 8e
torus, 21
transverse section, 61
Tremandra, 96
TREMANDRACEAE, 96
Trifolium, 120
 repens, Pl. 7e
trigger-plant, Fig. 100
 grass, Pl. 9h
trigger-plants, 191
Tristania, 239
trivial name, 54
tubers, 43
tulip, 219
Tulipa, 219
type genus, 56

Ulex europaeus, stamens, 13

vacuoles, 42
variety, in classification, 54
vascular system, 50
Velleia paradoxa, 189
 spur, 9
velvet-bush, 147
Venus fly-trap, leaves, 44
Vicia faba, 120
VISCACEAE, 87
Viscum album, 87
Vitis, tendrils, 43
von Mueller, Ferdinand, 55

waratah, 72
Watsonia, 228
wattle, 110

cape, 110
coast, Pl. 7c
cotyledons, 37
Flinders Range, Fig. 50
gold-dust, Fig. 48
golden, Fig. 51
gynoecium, 15
hedge, 111
myrtle, Pl. 7b
spreading, Pl. 7a
sunshine, Fig. 52
white sallow, Fig. 49
winter, Fig. 50
seeds, germination, 39
wax-berry, 171
wax flower, 153
long-leaf, Fig. 65
websites, 262
wedding bush, 139, Fig. 70
Westringia, 185
wheat, 211
whorl, of flower parts, 7
willow myrtle, 153
willow, native, 98

wing (petal), 121, Fig. 57–8
winter cherry, calyx, 9
Wisteria, 120
stems, 43
Wolffia australiana, Pl. 11f
Woollsia, 171
pungens, Fig. 96
Wurmbea, 219
dioica, Fig. 121
unisexual flowers, 23

Xanthorrhoea, 210, 218
australis, Pl. 12h–i
XANTHORRHOEACEAE, 210, 218
Xerochrysum bracteatum, Fig. 104
xylem, 50

Zantedeschia aethiopica, Pl. 6b
Zieria, 129
arborescens, Figs. 67–8
Zinnia, 193
zygomorphy, 11
zygote, 36